SOTHIS PRESS

Sothis Press aims to completely reimagine the public domain book with an emphasis on universal access to captivatingly designed print editions. We believe that technology will enable every single book ever published to be perpetually accessible in beautiful print form. Sothis is beauty in ubiquity.

www.sothis.press
1 Waterhouse Square, London, ECIN 2ST

The Indo-Chinese Forest Ox or Kouprey

Harold Jefferson Coolidge Jr.

SOTHIS PRESS

ISBN 978-1-0365-2437-1

This edition is published by
Sothis Press © 2023. The company
is registered in England and Wales.
The Company Number is 14900389.

Memoirs of the Museum of Comparative Zoölogy
AT HARVARD COLLEGE
Vol. LIV. No. 6

THE INDO-CHINESE FOREST OX OR KOUPREY

BY

HAROLD JEFFERSON COOLIDGE, JR.

With Eleven Plates

CAMBRIDGE, U. S, A.
Printed for the Museum
1940

TABLE OF CONTENTS

TABLE OF CONTENTS

THE INDO-CHINESE FOREST OX OR KOUPREY[1]

INTRODUCTION

In June 1937 Professor A. Urbain, director of the Vincennes Zoological Garden, first published in the Bulletin of the Zoological Society of France the description of the "Kou Prey" or wild ox of Cambodia which he names "Bos (Bibos) sauveli." A number of short papers on this rare animal have subsequently appeared, most of them describing externally a young male that is living at the Vincennes Zoo and is now in his fourth year. This living animal was designated as the holotype of the new species in a report of the French Academy of Sciences, Dec. 27, 1939.

During the winter of 1938–39 the "VIIe Expedition en Indochine" led by Mr. Jean Delacour had the good fortune to collect a fine specimen of this rare ox locally known as the "Kouprey" or "Kouproh." The animal was shot by Mr. François Edmond-Blanc accompanied by the hunting guide A. V. Pietri. This rare specimen, an old adult bull, was presented to the Museum of Comparative Zoölogy by James Cowan Greenway, Jr., who was a member of the expedition, and has been for some years Associate Curator of Birds on the staff of this museum.

As far as I can determine this is the first specimen of the kouprey that has been available for comparison and detailed description in any natural history museum. The present memoir deals with this most remarkable bovid.

The history of our ten-year knowledge of the animal is detailed in the Historical Appendix.

In the "comparative description" the kouprey is compared with a gaur and bantin shot within 200 km. of the same locality, for before the discovery of the kouprey these were the only two living kinds of wild taurine cattle from the forests of southeastern Asia (I regard the gayal as a semi-domesticated descendant of the gaur).

In my external comparison of these three forms I have described the kouprey hide in considerable detail. The principal significant external differences by which the kouprey differs from the two others are limited to the marking of the lower legs, the peculiar horns with their relation to the skull, and the elongated tail.

On comparing the measurements of the skull and skeleton of the kouprey with those of a gaur and a bantin, the kouprey revealed important differences.

[1] Published with the aid of a special gift from Mr. George R. Agassiz.

Some of the points suggest a possible relationship or parallelism in development to the wild yak, others to the zebu, and still others to the Indian water buffalo.

In comparing the kouprey skull directly with allied living bovid genera the greatest resemblance is to the zebu (Bos indicus) but there are important differences of kind between the two forms, especially in the teeth, horn-cores, and occipital region. The kouprey skull, while clearly that of an Asiatic taurine, shows more primitive features than any of the other surviving forms.

The comparison with allied fossil genera is made possible on account of Dr. Guy E. Pilgrim's comprehensive monograph on "The Fossil Bovidae of India" (Calcutta, 1939). Parts of this paper have been extensively quoted and Pilgrim's views on the phylogeny and classification of the Bovinae have been heavily depended upon. According to the scales set up by Pilgrim to determine primitive and progressive characters, the kouprey qualifies as an unusually primitive bovid resembling in some ways Proleptobos. When it came to the examination of the horn-core I found it unlike any form recorded by Pilgrim and quite different from that of any of the living Bovinae. Pilgrim considers the retention of primary keels as a primitive feature and our kouprey skull has a very fine posterior keel for almost two-thirds of the length of the horn-core. In order to determine whether this could possibly be an aberrant inner keel instead of a primary posterior keel, Dr. E. H. Colbert of the American Museum was consulted. He agrees that this is probably a primary keel. The horn-core of the younger kouprey from the American Museum's frontlet shows a deep double-grooved posterior ridge in the exact place where the keel should be, which appears to be an early stage in the same development. The most probable time for the separation of this genus from the main Proleptobos-Taurina stem is indicated on a modified part of Pilgrim's phylogenetic chart (see fig. 9, p. 494).

In this memoir I have adopted the following classification slightly modified from Schwarz (1937) and Pilgrim (1939):

Family—Bovidae

Subfamily—Bovinae

Generic Groups	*Genera*
Bubalina:	Parabos, Proamphibos, Hemibos, *Anoa,** *Bubalus, Bucapra*
Syncerina:	*Syncerus*
Leptobovina:	Proleptobos, Leptobos
Taurina:	*Novibos***, *Bibos, Poephagus, Bos, Bison,* Platybos.

*Living in italics.

**Described in this paper.

Our knowledge of the kouprey's distribution is extremely limited. The only definite localities are Tchep, the type locality, and Samrong (Samerong) near Krati where our specimen was procured and where Pietri probably procured the two other specimens mentioned in the Historical Appendix, p. 517. The localities are shown on the accompanying map (fig. 1.), and the fact that there are

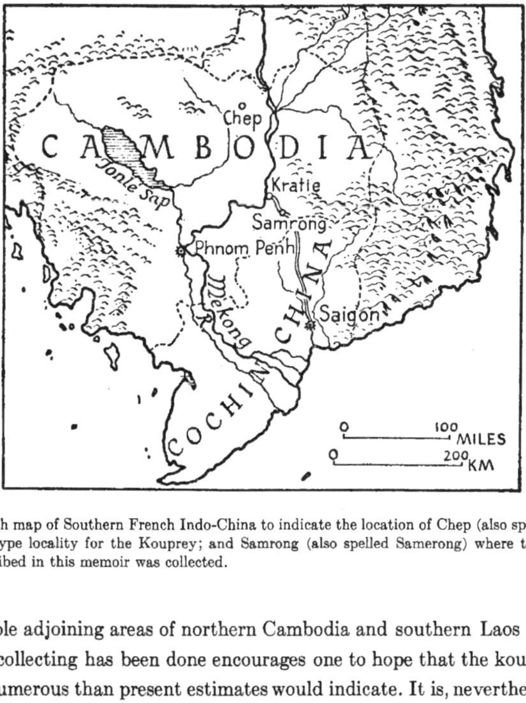

Fig. 1. Sketch map of Southern French Indo-China to indicate the location of Chep (also spelled Tchep) the type locality for the Kouprey; and Samrong (also spelled Samerong) where the specimen described in this memoir was collected.

considerable adjoining areas of northern Cambodia and southern Laos where no big-game collecting has been done encourages one to hope that the kouprey may be more numerous than present estimates would indicate. It is, nevertheless, most essential that the Government of French Indo-China should immediately recognize the importance of making every effort to preserve this interesting and rare primitive wild bovid, and especially to protect it against meat or trophy hunters and live-animal dealers. It is also important that if possible more detailed scientific information should be procured about it. Notes on the kouprey's life history would be of special interest, and such a study might well indicate the reason for the fraying of the tips of the horns. The discovery of the kouprey, a new genus of bovid, in Cambodia in French Indo-China in 1937 is quite as unexpected as Sir Harry Johnston's discovery of the Okapi in 1900 in the Ituri Forest of the Belgian Congo. The finding of the kouprey is also of special significance to the palaeontologists because of its relationship to the fossil bovids, especially of the Siwalik fauna, and to the archaeologists and zoologists who are searching for possible living representatives of the probable ancestors of neolithic domesticated cattle.

I wish to acknowledge with thanks the loan of a kouprey frontlet and other comparative material from the American Museum of Natural History, the loan of a wild yak skull from the Academy of Natural Sciences in Philadelphia, photographs and information from the New York Zoological Society, publications and information from the U. S. Bureau of Animal Industry, technical assistance from the Anatomy Department of the Harvard Medical School. I owe special thanks to J. C. Greenway, Jr., who procured this valuable specimen for us and has helped in its study by obtaining additional information concerning it. I am also indebted to M. François Edmond-Blanc for shooting such a fine specimen, to Thomas Barbour, Glover M. Allen, A. S. Romer, L. Griscom, Donald Carter, Harold E. Anthony, E. C. Colbert, W. H. Black, William F. Ross, Mrs. Frederick S. Goodwin, Dr. A. Wetherford, and Dr. S. C. Sossman for assistance in various ways in this study of the kouprey.

Mr. Eugene N. Fischer has shown his usual skill and accuracy in his drawings of the kouprey skull, and especially in the colored plate which he has prepared.

Miss Jesse H. Sawyer has also devoted many hours to the preparation of excellent drawings, and Mrs. Charles Meyer deserves special thanks for her patient helpfulness with the preparation of the manuscript and tables as well as her drawings of crown patterns of bovine teeth.

TAXONOMY

NOVIBOS, genus novum

Diagnosis.

Bovinae; size and external markings like Bibos; tail long; lower foreleg long black longitudinal stripe; horns large with yak-like curve,—near tips heavily frayed around entire circumference; skull narrow; palatal branch of premaxilla long; projection of premaxilla beyond nasals short; frontal short; tooth row short; orbit close to horn-core; bases of horn-cores approximated posteriorly; angle of horn divergence narrow; no intercornual ridge; horn-core section at base a flattened oval with prominent primary posterior keel; parietal sloping gradually toward occiput as in Bubalus; infracristal occiput high and subtriangular with prominent crests; space between occipital openings of temporal fossae narrow; upper pm^4 antero-posteriorly compressed; upper molars quadrate; a marked triangular ethmoid vacuity.

Description.

Bovinae, size between bantin and gaur; hide blackish brown; coat short glossy hairs; lower legs white stockings with dark stripe down middle of foreleg; dark hairs above hind hoofs; in old male white dorsal stripe on midline of lower back; marked dorsal ridge as in gayal; long tail reaching below hocks, tip bushy; ears small and narrow; legs and hoofs slender; face elongated and well-rounded, profile of forehead not concave; large dewlap.

Horns large, curving backward, outward and upward, then forward, and finally upward and inward with slight backward inclination as in wild yak; hornsheath dark, flattened at base, with corrugations as in a buffalo, changing to light smooth-surfaced oval shape, terminal third rounded; black tips greatly frayed forming collar of shredded horn, 150 mm. below tip; horns close together at base, horns longest and spread widest of living wild Bovinae except buffaloes.

Skull length close to bantin, but narrow zygomatic, frontal, intercornual, palatal, braincase widths; palatal part of premaxilla long; frontals short; projection of premaxilla beyond nasal short; height from palate to posterior tip of nasals shortened; general shape occiput narrow and square; occipital crest prominent and subtriangular; posterior ends of temporal fossae narrowed, approximating each other closely on occiput; no intercornual ridge; no frontal eminence, frontal and parietal have continuous gradual slope toward occiput and horn-core bases; parietal covers slightly less than ½ top of braincase; horn-cores have smallest angle of divergence (105°–125°) of male living wild bovinae except African buffalo and anoa; basal third of horn-core has very flattened oval cross-section; in old male marked primary posterior keel 9 mm. wide extending out two-thirds of length from base; tip end of horn-core directed anteriorly; maxillary tooth row short; length of combined molars short; combined premolars short; upper PM^4 greatly compressed, upper M^1, M^2, and M^3 quadrate; orbit close to base of horn core closer only in Bubalus; center of palate projects more than 30 mm. behind tooth row as in some Bos, Leptobos, and Bubalus; premaxilla contacts nasals; anterior nasal spines two, central; paroccipital heavy, ridged; foramen magnum small; lachrymal not deepened anteriorly; ethmoid vacuities triangular, prominent; scapula narrow; metapodials proportionately long and slender; radius lengthened in proportion to humerus; shortened ventral length of fused sacral vertebrae.

Genotype. Bos (Bibos)sauveli Urbain (1937, 1939)

Hypotype. Novibos sauveli, M.C.Z. 38108

NOVIBOS SAUVELI (Urbain)

Plate I

1937 Le Kou-Prey, etc. Ach. Urbain, Bull. Soc. Zool. de France, pp. 305–307.

1937 Le Kou-Prey, etc. Ach. Urbain, Mammalia, pp. 256–258.

1939 La collection des Bovines asiatiques du Parc Zoologique du Bois de Vincennes. Ach. Urbain, P. Rode et M-A. Pasquier.

1939 Note Bos (Bibos) sauveli. Ach. Urbain, Bull. du Muséum, XI, No. 6.

Diagnosis. This being the only species known, the diagnosis is that of the genus.

Holotype. A living male 4 years old Dec. 1939 living in the Vincennes Zoological Garden (Paris) and figured in Urbain's report (1939). (see Plate III fig. 1).

Hypotype. Old adult male skin, skull, and part skeleton M.C.Z. 38108 from near Samrong[1], Province of Krati, Cambodia, French Indo-China, March 16, 1939. First described in this report. (see Plate II).

Collector. F. Edmond-Blanc and A. V. Pietri.

Referred specimen. Adult male frontlet and horns A.M.N.H. 89003 from Province of Krati, Cambodia, French Indo-China, Feb. 1939. Collector, Ezra Cornell and A. V. Pietri. (see Plate III fig. 3 and Plate V).

Remarks. For description see following pages 427 to 469.

Novibos sauveli M.C.Z. 38108 has been designated as a hypotype because many of the characters of the genus are most marked in this first-described skel-

[1] Also spelt Samerong.

eton of an old adult wild male. The fact that the holotype is still a young animal reared in captivity from an early age may in some measure affect the normal growth of its skeleton a condition often noted in zoo animals. Frizzell (1933, p. 653) defines a hypotype as "a described or figured specimen used in publication in extending or correcting the knowledge of a previously defined species."

COMPARATIVE DESCRIPTION

HIDE

The reported field measurements of this wild ox were:

Total length—235 cm.
Height from fore-foot to shoulder—171 cm. (5′ 7½″)
Circumference around the chest—210 cm.
Ear—17 cm. long and 12 cm. wide.
From line between horns to tip of nose—55 cm.
Neck circumference—170 cm; behind horns—30 cm.
Perpendicular length of dewlap—44 cm.

These are more comparable in general size with the gaur measurements than with those of bantin, though a large bull bantin of the Malayan race has been recorded with a shoulder height of 1633 mm.

The tail length (measured on tanned skin) was 1025 mm. of this wild ox as compared with 660 mm. (tanned skin) for an adult female gaur and 740 mm. (tanned skin) for an adult bull bantin, all from French Indo-China.

The head gives the general appearance of being brownish black (see Plate I).

The general color of the short sleek shiny hair that covers the head and shoulders is dark brown tending to black.

The muzzle is slate black with a dark chestnut-brown[1] area about 35 mm. wide on either side connected by an irregular patch of similar color about 20 mm. wide above the muzzle. The upper lips have white hairs in a band about 12 mm. wide at the muzzle and narrowing as they extend for 100 mm. on either side of the muzzle. The hair on the lower lip is a sepia color with a few scanty white hairs in a patch about 30 mm. long on either side.

The face below the level of the eyes and the neck is a blackish mummy brown. The forehead from between the eyes to the horncrest is a chestnut brown. An area of the same color about 30 mm. wide surrounds each eye. There are

[1] Note color terminology from Ridgway, 1912. Osteological terminology from Sisson, 1917.

white hairs inside of the slender ears, some of them rather long. The backs of the ears are scantily haired and the color is the same glossy blackish mummy brown which uniformly covers the shoulders, neck and large dewlap, and which is replaced by a deep olive gray in the area behind the shoulders and back of the dorsal ridge. This change of color is probably due to the fact that the darker hairs have been rubbed off certain areas of the skin and the dark skin is covered with very fine pale hairs. There is a curious mottled effect in the area where this transition takes place. Between the upper forelegs and below the dewlap the hair color is lighter, close to mars brown mixed with short gray hairs. The face, neck and shoulders are furrowed with heavy wrinkles most of which extend up and down perpendicular to the horizontal (see plates I and II).

The upper foreleg above the carpus is the same blackish mummy brown as the neck. Covering the carpus and extending down the front of the leg to the hoof is a band of 10 to 15 mm. long blackish mummy-brown hairs sprinkled with occasional white hairs. This dark area is 80 mm. wide at the carpus and narrows down to 30 mm. wide until it reaches the phalanges where it again widens to 70 and 80 mm. The sides and back of the lower leg are covered with short white hairs which form a marked contrast to the anterior stripe.

Deep olive gray, with fine gray hairs where the skin has been greatly rubbed, covers the sides of the barrel and extends up to within 100 mm. of the center line of the back. Along the very center of the back there are traces of a whitish stripe extending from the back of the dorsal ridge for 700 mm. to a point within 200 mm. of the base of the tail. This stripe where it is most marked is made up of completely cream-colored hairs and is 12 mm. wide. On either side of it there is a mottled area about 100 mm. wide with blackish mummy-brown hairs spotted with patches of olive-gray skin and fine gray hairs.

The under side of the belly from the forelegs back for 700 mm. is a light mars brown. There is a small urethral tuft made up of gray and brownish hairs about 30 mm. long. The skin around the scrotum is light flesh-colored with spots of dark pigment. The scrotum is darkly pigmented with areas of very fine light-colored hairs. In front of the scrotum are four rudimentary teats.

The hips, thigh, and upper hind leg have the characteristic blackish mummy-brown hairs bordered in front by the same mottled effect already referred to. From the knees down to the fetlock joint, about 280 mm., the lower leg is covered with short white hairs. From this point for the 160 mm. down to the hoof there is a mixture of white, black and brown hairs about 6 mm. long that gives a general gray effect to the foot coloring.

The coloring of the tail is mottled with blackish mummy-brown hairs and areas of fine light-gray hairs extending for some 370 mm. back from the base of the tail. For the next 230 mm. the tail is mummy brown above with a generous sprinkling of white hairs the rest of the way around it. The long hairs of the brush on the end of the tail start at about 580 mm. from the base of the tail and cover 430 mm. of the terminal part of the tail. The upper fifth of the brush is made up of long black and long white hairs; the next two-fifths almost completely of white hairs; and the last two-fifths of predominantly black hairs with a sprinkling of white. The hairs of the fairly dense brushy tail average about 120 mm. in length.

The tips of the horns, extending for 210 mm. on the right side and 120 mm. on the left side out from the collars of shredded horn characteristic of this animal, are primarily black. From the shredded collars for about 450 mm. down (outside curve) the horn color is a deep dull brownish yellow-green; the ridged basal parts of the horns which extend for 180 mm. are predominantly black with a narrow band of yellowish or deep dull yellow-green horn color on the front and back faces of the flat oval, which is the cross-section shape of the horns at their base. The shape of the horns will be described in the section dealing with the skull.

The hoofs are black and the hind ones unusually slender. The greatest length and breadth of the horny casing of the hind hoof is 109 mm. by 40 mm. as compared with 124 mm. by 51 mm. in the fore foot. The differences in the shape of the hind hoof of kouprey, gaur, and bantin are shown in the accompanying diagram (see fig. 2, next page).

Professor Urbain (1937) describes the gray ox as an animal that differs from the bantin by its great height, certain ones reaching 1.90 m. at the shoulder. He notes that the hide of the kouprey is entirely gray in the females and young animals, and a beautiful dead black in the old bulls, with grayish patches on the shoulders and hips. These animals, he said, had a very pronounced dewlap and the white on the legs extended far up. The horns are cylindrical, widely spread, recurved forward in the bulls and lyre-shaped in the calf. They possess sometimes, especially in the old males, a few cm. from their ends, a very curious little collar made of external pieces of horn that they shred off as a result of repeatedly digging in the ground which he says the animals frequently do. The nape is narrow, covered with short velvety hairs. The withers are powerful without muscular deformation and prolonged behind on the dorsal region. The ears are light and finely tapered; the tail long. The legs are longer than in the bantin and very slender. The feet are very slender and of an astonishing lightness. Altogether, he says, the "Kou Prey" is a graceful animal whose external morphology is entirely different from that of gaurs and bantins to which certain authors wanted to have them belong.

The description of the hide of our specimen and the colored plate based on photographs and field measurements conform in general to those given by Prof. Urbain and indicate that our kouprey belongs to the form which he described as "Bos (Bibos) Sauveli."

a *b* *c*

Fig. 2. Side and basal view of hind hooves (¼ nat. size) of *a*. Kouprey (M.C.Z. 38108); *b*. Gaur (M.C.Z. 36778); *c*. Bantin (M.C.Z. 36669).

In external appearance (without discussing here the shape of the horns which is treated in the section on the skull) our kouprey[1], or specimen of Asiatic wild ox, has a number of characters that are common to both the gaur and the bantin and certain others that are characteristic of the gaur or the bantin or the semi-domesticated gayal. There are other characters in which it differs greatly from all three in some cases and resembles the yak or the Indian water buffalo. Differences are chiefly in the skull, but also the shape of the horns, the curious shredding of the horn sheath close to the tips, the greater length and greater bushiness of the end of the tail, certain skeletal differences, the dorsal and leg stripes, and the more elongated but less tapered face which is suggestive of some races of domestic cattle rather than the features of a wild bantin or gaur.

The kouprey suggests an intermediate form between the gaur and the bantin in its size. The height of our specimen does not exceed that of a large bull bantin of the typical Javan race (Bibos banteng banteng) although it does exceed by a few centimeters records for the Indo-China race, Bibos banteng subsp.? The external measurements come well within the range of those for the gaur, although

[1] I have adopted the single word spelling used by Mr. Omar Sarraut. Other spellings that have been used are KouPrey, Kou Pray, Kou proh and Kou-Prey.

a large adult gaur stands 254 mm. higher at the withers and would be 457 mm. longer from nose to rump, altogether a larger animal. The adult kouprey appears to have a lighter build than the gaur and a heavier one than the bantin.

The hoofs of the kouprey although slightly more slender are quite similar to those of the gaur and the bantin. The horn coloring is similar to that of the gaur and bantin. The general blackish-brown and grayish coloring of hide and hair is more characteristic of the gaur than of the bantin, but old bulls of the typical Javan race of bantin are occasionally blackish brown or black. The generally white lower legs of the kouprey below the hocks are characteristic of both gaur and bantin, as are the white lips, and long white hairs in the ears. The slightly lighter-colored area above the muzzle suggests in its location the white band characteristic of bantins, and the tawny band round the lower part of the jaw immediately above the muzzle is found in certain races of gaur. The nostril of the kouprey has a prominent upper wing which makes the nostril more conspicuously notched (see plate II) than is the case in the gaur. The bantin is somewhat more similar to the kouprey in this respect.

The kouprey has a slightly light-colored area on the forehead extending down to about the eye level. This corresponds with the frontal area in the gaur that is generally covered with long tawny hair and in the bantin this same area is generally whitish or lighter than the rest of the face except for the lips and bands around the eyes and muzzle.

The kouprey resembles the bantin in having no long hairs on the throat; the hide (except where it is rubbed off) is covered with short glossy hair except for the tail, inside of ears, front of lower legs and small urethral tuft. The white 'stockings' on the lower leg are not golden brown at point of origin or all white as in the gaur, but they are similar to the clear white of the bantin except for the dark stripe in the middle of the forelegs and the mixture of dark hairs in the foot above the hoof of the hind legs. Bantins have an area of light color, often white, about each eye. The kouprey has a dark chestnut area about its eyes that is little lighter than the surrounding hair but suggests this characteristic marking of the bantin. In the chestnut color of the eye itself, the kouprey is similar to the bantin. That of the gaur is blue.

The shape of the head is elongated more as in the bantin[1] than the gaur with

[1] The skull of the female Bibos banteng banteng (M.C.Z. 13254) from Java resembles the kouprey more than the male in the fact that the skull is long and narrow, the small rounded horn-cores are directed almost directly backwards forming an angle of divergence on the frontal of about 80°. The large temporal fossae approach each other on the occiput, the interval between them being 76 mm. There is no intercornual ridge but there is a marked pointed frontal eminence.

the absence of the concave profile of the forehead so characteristic of the latter. In the close approximation of the bases of the horns it resembles more the male Bornean bantin (Bibos banteng lowi) than any described race of gaur.

In external appearance the kouprey resembles the gaur in its marked dorsal ridge which is not as prominent as in the typical gaur but which is not characteristic of the living bantin. This ridge extending back from the withers terminates in a less marked step than in the gaur. The kouprey is like the gaur in the absence of the light rump-patch which is a characteristic of most races of the bantin. The general blackish color of the body in old bulls is more characteristic of the gaur than of the bantin, and the bantin bulls in the Indo-China race (Bos banteng subsp.?) are usually a bright orange, tan, or fawn color with a black band encircling the fore-leg above the knee. A large dewlap is not generally characteristic of the wild Indian gaur but as Lydekker points out in his 'Game Animals of India' (1924, p. 61): "In Travancore some of the old bulls display a strongly developed dewlap, although this character is not constant in the district. More important still is a statement by Mr. C. W. A. Bruce to the effect that the Burmese gaur are always distinguished by the presence of a large dewlap, as well as by their very dark color." Theodore Hubback writing of the Malayan gaur or seladang (1938) says: "There appear to be two distinct phases of development of the dewlap in the seladang. In some cases mature seladang have a very pronounced dewlap; in other cases the dewlap is hardly developed at all. I have found that seladang with heavy dewlaps are generally taller than those with little or no dewlap, have longer horns and have more light hair around the muzzle. Those with little or no dewlap are generally stockily built, but less in shoulder measurement and although the light patch on the chin seems to be common to all seladang those without a developed dewlap seldom have any light patch over the nose. This applies to cows as well as bulls, but the dewlap is never well developed in the cow."

The gayal which is found in a domesticated or semi-domesticated condition both to the north and south of the Assam valley is according to Mr. E. C. Stuart Baker and Lydekker probably an artificial derivative from the gaur which it very much resembles. The bull gayal is smaller than the wild gaur, has a greatly developed dewlap, a straight line on the vertex of the skull between the horns in place of the arched ridge typical of the gaur, and a somewhat less elevation of the dorsal ridge.

In these points it resembles the kouprey more closely than the gaur does, but here the resemblance ends for the gayal has shorter limbs, a very wide short forehead, widely separated, slightly curved horns, and other characters which remove it from the kouprey.

Summary.

The external appearance of the kouprey agrees with the brief description given by Professor Urbain (1937). A comparison with the gaur and the bantin hide indicates a common resemblance or a possible intermediate form in matters of size, lip, ear, muzzle, and forehead markings, general body color, white stockings, shape of hoof, and horn color, and difference in the dark areas on the fronts of the lower legs, and the white dorsal stripe.

The kouprey resembles the bantin more than the gaur externally in its short glossy hairy coat and the distribution of areas with longer hair, in the white of its legs, in the lighter area around the eye, in its chestnut colored eyes, in the shape of its head, and in the position of its horns.

The kouprey resembles the gaur more than the bantin in its marked dorsal ridge, the dark general coloring, the dewlap, and in the absence of a light rump patch. There is a resemblance to the gayal in the large dewlap, the absence of the arched ridge on the vertex of the skull between the horns, and a less marked dorsal ridge than in the typical gaur. The kouprey differs from bantin, gaur, and gayal in basic skull characters including the shape and character of the horns, the greater total tail length, and greater bushiness at the tip of it, the conspicuously notched nostril, the smaller narrower ears, the more graceful build, slender legs, and the more elongated but less finely chiseled face, which would never be mistaken for a living gaur or bantin even in a poor photograph. The general rectangular build of the kouprey so well indicated by Professor Urbain's photographs of the young one in Paris is distinctly different from similar pictures of young gaur, bantin, and gayal (*Mammalia* (Paris), p. 124, pl. 9, 1939).

SKULL

The known living forms of wild oxen of the Indo-Malayan countries, with the exception of Novibos sauveli, the kouprey, only recently described by Urbain, have been the two well-known species, Bibos gaurus, the gaur; and Bibos banteng, the bantin; of these species a number of undoubtedly valid local races have been described.

In order to determine possible significant differences between the gaur, the bantin and the kouprey, similar measurements and in some cases scale drawings have been made of the skull of an adult male gaur (Bibos gaurus readi) and an adult male bantin (Bibos banteng subsp.?) collected in 1937 by Andrew Wylie close to Banmethuot in southern Annam only about 180 m. east of Samrong in Cambodia where our specimen of the kouprey was shot. Such a comparison of these three wild oxen should not only indicate which form the kouprey most closely resembles or in what respects it is an intermediate, but also it should point out certain differences or similarities between the gaur and the bantin not previously recorded.

Tables I and III show the comparative skull and horn measurements of the kouprey, bantin, and gaur. Tables II and IV show comparative measurements between parts of the kouprey skull and the frontlet loaned by the American Museum.

TABLE I

COMPARATIVE SKULL MEASUREMENTS AND CORRELATIONS WITH BASAL LENGTH RATIO

	Kouprey MCZ 38108 Actual Measurements	Bantin MCZ 36669 Actual Measurements	Gaur MCZ 36670 Actual Measurements	Bantin MCZ 36669 Corrected meas. A. and % deviation of actual meas. to Kouprey B.		Gaur MCZ 36670 Corrected meas. A. and % deviation of actual meas. to Kouprey B.	
LENGTHS				A.	B.	A.	B.
1. Greatest length	470	465	520	481	−3.3	496	+4.6
2. Basal length	446	456	470				
3. Palatal length	302	291	312	309	−5.9	318	−2.0
4. Palatal length of premaxilla	132	100	113	135	−26	138	−18.1
5. Basion to rear median point of palate	144	165	158	147	+11.3	152	+3.7
6. Nasal length	202	177	230	206	−14.1	213	+7.4
7. Length of temporal fossa	165	155	177	168.5	−8	174	+1.6
8. Occiput to anterior point of supraorbital process	210	220	260	215	+2.3	222	+14.7
9. Foramen magnum to ant. end max. tooth row	300	313	328	307	+1.9	316	+3.8
10. Max. tooth row length	118	134	156	120.5	+9.8	124.5	+20.1
11. Mandibular tooth row length		138	164				

	Kouprey MCZ 38108 Actual Measurements	Bantin MCZ 36669 Actual Measurements	Gaur MCZ 36670 Actual Measurements	Bantin MCZ 36669 Corrected meas. A. and % deviation of actual meas. to Kouprey B.		Gaur MCZ 36670 Corrected meas. A. and % deviation of actual meas. to Kouprey B.	
WIDTHS				A.	B.	A.	B.
12. Mandible: Horizontal distance front of tooth row to posterior angle		262	282				
13. Length of mandibular diastema		112	118				
14. Zygomatic width	194	209	234	198.5	+5.0	205	+12.5
15. Frontal width at supraorbital process	202	204	263	206	−1.2	213	+19
16. Greatest mastoid width	227	238	287	232	+2.3	239	+16.6
17. Braincase width	111	122	122	113.5	+7.0	117	+4.0
18. Palatal width	83	90	98	85	+5.6	87.5	+10.7
19. Postorbital constriction width	170	187	242	174	+6.9	179	+26.0
20. Width betw. supraorbital foramina.	99	104	153	101	+2.7	104	+32.7
21. Max. width at upper M1.	129	132	148	132	0.0	136	+8.1
22. Width betw. bases of horn-cores (anterior)	190	273	342	194	+141.0	200	+171.0
23. Same (posterior)	94	215	255	96	+113.0	99	+157.0
24. Length of intercornual ridge	118	285	305	120.5	+142.4	124	+146.0
HEIGHT							
25. From palate to upper tip of nasals	133	147	173	136	+7.5	140	+19.0
26. Foramen magnum to frontal eminence	151	176	184	154.5	+11.4	159.0	+15.5
27. Foramen magnum to occipital crest	84	82	88	85.5	−4.0	88.5	−0.6
MISC. MEASUREMENTS							
28. Greatest combined nasal width	66	57	74	67.5	−15.6	69.6	+6.0
29. Occipital condylar width	106	108	121	108.5	−0.4	112	+7.5
30. Foramen magnum width	35	40	40	36	+10.0	36.9	+7.7
31. Width of mandibular condyle		49	57				
32. Transverse diameter left orbit	58	62	62	54.3	+4.4	61.2	+1.3
33. Antero-posterior diameter left orbit	58	67	64	59.3	+11.4	61.2	+4.4
34. Shortest width betw. bases of horns	75	215	270	77	+292.0	79	+342.0

DIAGRAM OF PERCENTAGES FROM TABLE I

Gaur ■, Bantin ▨, Kouprey vertical line

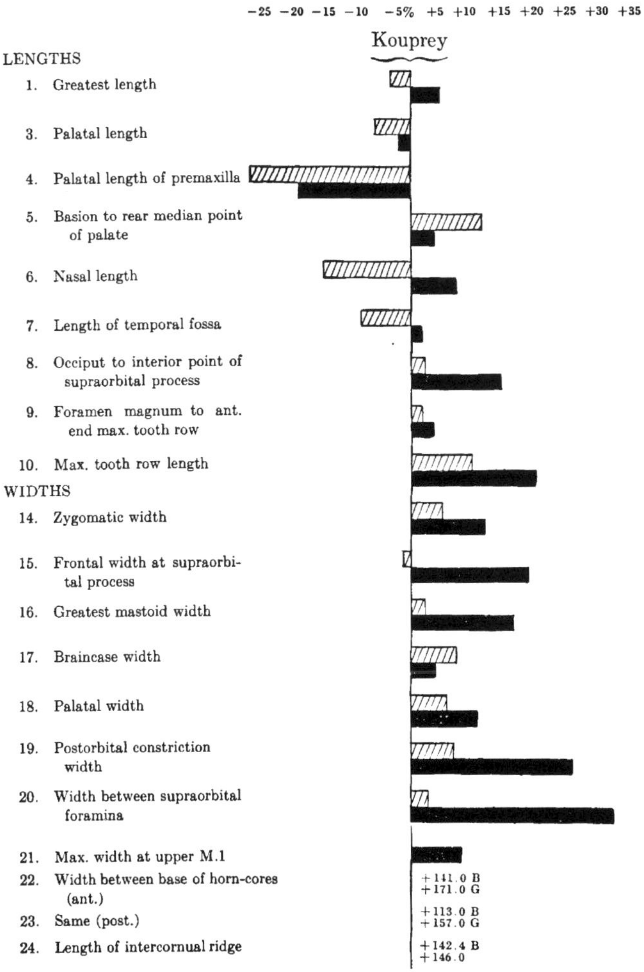

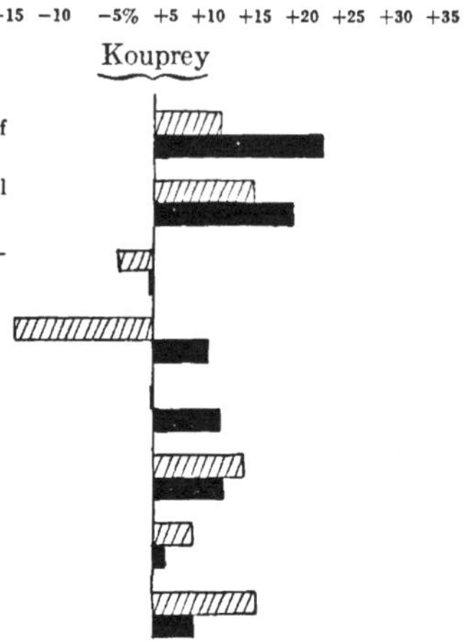

TABLE II

KOUPREY MEASUREMENTS

SKULL

	Kouprey Frontlet A.M.N.H. 89003	Kouprey Skull M.C.Z. 38108
1. Orbital rim to base of horn-core	80	75
2. Orbital rim to top of occipital crest	191	185
3. Width between posterior part of orbits	200	200
4. Width of post-orbital constriction	180	170
5. Width between supra-orbital foramina	90	99
6. Width on occiput between openings of temporal fossae	78	75
7. Height from external occipital protuberance to top of parietal	61	68
8. Width between orbital foramina	98	100
9. Greatest width of braincase	110	111
10. Greatest length of braincase (inside)	120	—
11. Greatest length of temporal fossa (from rim)	123	135
12. Thickness of posterior wall of occiput	28	—
13. Angle of frontal and parietal	113°	118°
14. Angle of parietal and top of occiput	28°	30°

Note: The frontal of the frontlet is flat with the central suture slightly spread starting 85 mm. below the top of the parietal and extending for 85 mm.

ANALYSIS OF MEASUREMENTS

Because of the fact that the adult male gaur skull (M.C.Z. 36670) is somewhat larger than that of the kouprey and the adult bantin skull (M.C.Z. 36669) is slightly larger, a comparison of percentages based on corrected values indicates possible significant differences more readily than a comparison of the tabulated absolute measurements also recorded.

The measurements have been corrected according to their basal-length ratio using the measurements of the kouprey as a standard for comparison. The actual values divided by the corrected values give us in each case the percentage of deviation in both the bantin and the gaur from their corrected values, which have been derived by comparison with similar measurements of the kouprey. These values are not only recorded in Tables I and III but the percentages are shown in the block diagrams following the tables.

Skull Lengths.

1. In corrected greatest length the gaur exceeds the kouprey by 4.6%, while the bantin is 3.3% shorter. The kouprey is an intermediate in respect to this measurement, slightly closer to the bantin than to the gaur.
2. In basal length the bantin exceeds the kouprey by about 2% and the gaur exceeds the kouprey by about 5.5%. In this measurement the kouprey represents an extreme on the short side, closer to bantin than to gaur. The basal length is made up of the sum of number 3 and number 5 measurements in Table I which are separately compared.
3. In corrected palatal length the kouprey has the longest palate, exceeding both gaur and bantin but coming closer to the gaur.
4. In corrected palatal length of premaxilla the kouprey is greatest and closer to the gaur than the bantin which is 26% shorter in this measurement.
5. In corrected length from basion to the rear median point of palate, in Table I the kouprey is shortest and closest to the gaur which in this measurement approaches the kouprey by a difference of 3.7% compared with 11.3% for the bantin.
6. In corrected greatest nasal length the gaur exceeds the kouprey by 7.4% while the bantin is 14.1% shorter. The kouprey is an intermediate, closer to the gaur than to the bantin.
7. In corrected length of temporal fossa the gaur slightly exceeds the kouprey while the bantin is 8% shorter.

8. In the corrected length measurement from the occiput to the anterior point of the supraorbital process the kouprey is shortest and considerably closer to the bantin than to the gaur.

9. In the corrected length measurement from the foramen magnum to the anterior end of maxillary tooth row, the kouprey is shortest and closer to the bantin than to the gaur.

10. In the corrected length of the maxillary tooth row the kouprey is notably shortest and closer to the bantin than to the gaur. The maxillary teeth are crowded and worn in the kouprey as one would expect in an older animal. The kouprey teeth are much more quadrate than those of the gaur or bantin as will be noted later.

12. In the corrected mandibular length or horizontal distance from front of tooth row to posterior angle the gaur exceeds the bantin by 3.8%.

13. In the corrected length of the mandibular distance measured between PM^2 and C the gaur is shorter than the bantin by 6.4%.

In the thirteen corrected length measurements just compared, the kouprey is greatest only in palatal length and length of the palatal premaxilla, and least in basal length, basion to palate, occiput to supraorbital process, foramen magnum to PM^2 and length of maxillary tooth row. It is intermediate between the gaur and the bantin in greatest length, nasal length, length of temporal fossa.

The kouprey's palatal length is 2% greater than the gaur and 5.9% greater than the bantin's. The palatal portion and process of the premaxilla is 18 and 26% longer in the kouprey than the gaur or bantin. If we eliminate this from our total palatal length there remains the median length of the palatine process of the maxilla plus the horizontal part of the palatine bone and in this measurement the bantin exceeds the kouprey by 9% and the gaur exceeds by 11%. Therefore in the median line of the palate the kouprey shows from 18 to 26% greater length in the premaxillary part of the palate, and a 9 to 11% shortness in the rest of the palate compared with the bantin or gaur.

Summary.

On the basis of a comparison of the thirteen skull-length measurements the kouprey shows a difference of between 2 and 5% from the gaur or bantin in almost every one. It is extreme in length of the palatal part of the premaxilla and in the shortness of basal length, basion to palate, occiput to supra-orbital process, and length of maxillary tooth row. It is closer to the gaur than to the bantin in palatal length, also in nasal length, and in length of temporal fossa. Although somewhat intermediate in these last three. It is closer to the bantin than the gaur is in basal length, in frontal length, and in length of maxillary tooth row, and also in greatest length, although intermediate in this last measurement.

There appears to be a from 2 to 5% length difference between the gaur and the bantin. The gaur exceeds in greatest length, mandible length, length of maxillary tooth row, and is much greater in frontal length, while the bantin differs markedly in the very much shorter nasals, in the shorter temporal fossa, and in the shorter mandibular tooth row.

Widths.

14. In corrected zygomatic width the kouprey is exceeded by 5% in the bantin and 12.5% in the gaur.

15. In corrected frontal width at supraorbital process the bantin is 1.2% less than the kouprey and the latter 19% less than the gaur.

20. In corrected width between supraorbital foramina the kouprey is exceeded by 2.7% in the bantin and 32.7% in the gaur.

19. In corrected width of postorbital constriction we should expect to find somewhat the same relationship as in frontal width. The kouprey is exceeded by 6.9% in the bantin and 26% in the gaur.

16. In corrected greatest mastoid width the kouprey is exceeded by 2.3% in the bantin and 16.6% in the gaur.

17. In corrected braincase width the kouprey is exceeded by 4% in the gaur and 7% in the bantin.

18. In corrected palatal width the kouprey is exceeded by 5.6% in the bantin and 10.7% in the gaur.

21. In the corrected maxillary width at M^1 the kouprey and the bantin are the same and are exceeded by 8.1% in the gaur.

22. In corrected width between anterior bases of horn-cores the kouprey is exceeded by 141% in the bantin and 171% in the gaur.

23. In the same measurement as 22, only taken posteriorly, the kouprey is exceeded by 113% in the bantin and 157% in the gaur.

24. In corrected length of intercornual ridge the kouprey is exceeded by 142.4% in the bantin and 146% in the gaur.

Summary. Out of the eleven corrected width measurements involving the complete skull as distinguished from the width of individual bones or foramina which will be treated separately, the kouprey was the narrowest in nine out of these by a 2 to 5% difference in 3, and a 5 to 10% difference in 3, and a greater than 10% difference in 3.

In maxillary width the kouprey and bantin were the same and considerably narrower than the gaur. In corrected frontal width measured at the supraorbital process, the kouprey exceeded the bantin by only 1.2%, which could easily be accounted for by the slightly more prominent supraorbital bulge in the kouprey compared with the bantin. Both of them were exceeded in this measurement by 19% or more in the gaur.

Comparison of the corrected width measurements shows that in all eleven width measurements the kouprey is closer to the bantin than to the gaur except in width of braincase, where the kouprey is narrowest, exceeded by 4% in the gaur and 7% in the bantin. In seven out of the ten width measurements in which the kouprey is closer to the bantin, the bantin is closer to the kouprey than it is to the gaur. In the case of the palatal width the bantin is almost an exact intermediate between the kouprey and the gaur. In the width between the base of the horn-cores measured posteriorly and in the length of the intercornual ridge, the bantin and gaur were closer to each other than they were to the kouprey which was very much less. The gaur exceeds the bantin in all eleven width measurements with the exception of the width of the braincase already mentioned. The difference in all ten cases is greater than 5% and in several cases greater than 15%.

Heights.

25. In corrected height from palate to upper tip of nasals the kouprey is exceeded by 7.5% in the bantin and 19% in the gaur.

26. In corrected height from foramen magnum to frontal eminence the kouprey is exceeded by 11.4% in the bantin and 15.5% in the gaur.

27. In corrected height from foramen magnum to occipital crest the bantin is 4% less than the kouprey and the gaur 0.6% less.

Out of three corrected skull heights the kouprey is least by a difference ranging from 7 to 19% in two and greatest in one. In the height from palate to nasals the kouprey is least and closer to the bantin than the bantin is to the gaur. In the height of nuchal shield the kouprey is least and the bantin and gaur are close to each other, the bantin being only 4.1% nearer to the kouprey than to the gaur. In the height of the occipital crest from the foramen magnum the gaur and kouprey are very close to each other and exceed the bantin by about 4%.

Miscellaneous Measurements.

28. In corrected greatest combined-nasal width the kouprey is intermediate, exceeding the bantin by 15.6%. The kouprey is exceeded by 6% in the gaur.
29. In corrected occipital-condylar width the kouprey only exceeds the bantin by 0.4% but the gaur exceeds the kouprey by 7.5%.
30. In corrected foramen-magnum width the kouprey is exceeded by 10% in the bantin and 7.7% in the gaur.
32. In corrected transverse diameter of the left orbit the kouprey is exceeded by 4.4% in the bantin and 1.3% in the gaur.
33. In corrected antero-posterior diameter of the left orbit the kouprey is exceeded by 4.4% in the gaur and 11.4% in the bantin.

In the above miscellaneous measurements the kouprey is intermediate between the bantin and gaur and closest to the latter in nasal width. It is virtually the same as the bantin in the occipital-condylar width, and both are less than the gaur. The kouprey is smallest and closer to the gaur than to the bantin in width of foramen magnum and in both diameters of the orbit. In these measurements the gaur is conspicuous on account of its much greater nasal width than the bantin, smaller foramen magnum, and smaller orbits.

TABLE III

COMPARATIVE HORN MEASUREMENTS AND CORRELATIONS WITH BASAL LENGTH RATIO

	Kouprey MCZ 38108 Actual Measurements	Bantin MCZ 36669 Actual Measurements	Gaur MCZ 36670 Actual Measurements	Bantin MCZ 36669 Corrected meas. A. and % deviation of actual meas. to Kouprey B.		Gaur MCZ 36670 Corrected meas. A. and % deviation of actual meas. to Kouprey B.	
				A	B	A	B
34. Shortest width between bases of horn-cores	75	215	270	77	+192	79	+242
35. Greatest spread	840	740	825	859	−13.8	885	−6.7
36. Tip to tip	460	422	600	470	−10.2	485	+19.2
37. Left horn: length on outside curve	810	552	630	828	−33.3	852	−26.0
38. Circumference at base of left horn	359	310	392	367	−15.4	378	+3.5
39. Circumference 100 mm. from tip of left horn	103	63	88	105.3	−40.2	108.5	−19.0

DIAGRAM OF PERCENTAGES FROM TABLE III

Gaur ■, Bantin ▨, Kouprey vertical line

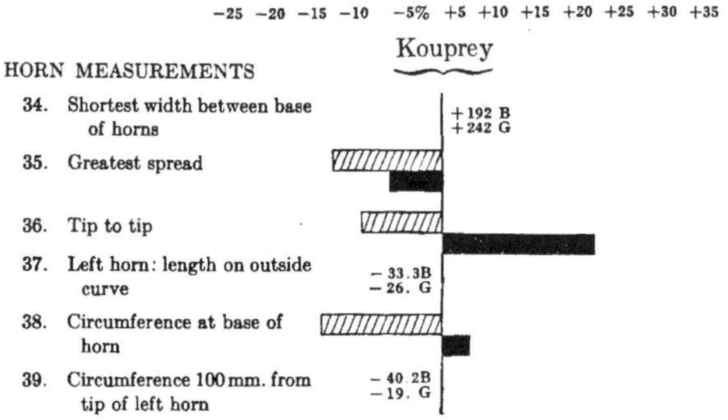

TABLE IV

KOUPREY HORN MEASUREMENTS

	Kouprey Frontlet A.M.N.H. 89003	Kouprey Skull M.C.Z. 38108
1. Tip to tip	420	460
2. Greatest spread	823	840
3. Left horn, fraying to tip	250	115
4. Right horn, fraying to tip	162	215
5. Narrowest width between horn bases posteriorly	122	75*
6. Greatest width between horn bases anteriorly	240	220
7. Outside curve—length of right horn	783	785
8. Outside curve—length of left horn	820	810
9. Outside curve—length of right horn-core	595	553
10. Circumference outside base of left horn-core	310	280
11. Thickness of horn sheath posteriorly, base of horn-core	10	23*
12. Thickness of horn sheath anteriorly, base of horn-core	6	7
13. Circumference 100 mm. from tip of left horn	104	103
14. Circumference 100 mm. from tip of left horn-core	130	140
15. Greatest width antero-posteriorly at base of horn (left)	75	70
16. Angle of horns divergence at frontal	125°	105°*

* Differences probably effected by greater age of specimen.

Horns.

Table III shows the comparative skull measurements of the kouprey, bantin and gaur. Table IV shows comparative measurements between the horns of the kouprey skull and those of the frontlet loaned by the American Museum.

34. In corrected shortest width between bases of horn-cores the kouprey is 192% less than the bantin and 242% less than the gaur.

35. In corrected greatest spread of horns the kouprey exceeds the gaur by 6.7% and the bantin by 13.8%.

36. In corrected distance between the tips of the horns the bantin is 10.2% less than the kouprey, and the gaur 19.2% more. One reason for this is that the horn tips of the kouprey are turned in.

37. In corrected length on the outside curve (left horn) the kouprey exceeds the gaur by 26% and the bantin by 33.3%.

38. In corrected circumference at base of horn the kouprey exceeds the bantin by 15.4% and is exceeded by 3.5% in the gaur.

39. In corrected circumference 100 mm. from the tip of the horn (left) the kouprey exceeds the gaur by 19% and the bantin by 40.2%.

Summary.

The horn measurements are highly variable although the horns of an adult bantin or gaur usually conform to an easily recognizable pattern of curve and shape which makes them easily distinguishable. The kouprey seems to have a definite yak-like pattern of its own. The horns are greater in spread and length than those of a large gaur, although close to the gaur in circumference at the base.

A comparison of the tabulated corrected horn measurements makes the kouprey closest to the gaur but significantly greater than the gaur or bantin in greatest spread, length of horn on outside curve, and in the horn circumference 100 mm. from the tip. The gaur is only 3.5% greater than the kouprey in the corrected circumference at the base of the horn, while the bantin is 15.4% less than the kouprey. The distance between the tips of the horns shows the kouprey to be intermediate in this very variable measurement with the bantin 10.2% less and the gaur 19.2% more. The most striking difference is shown in the shortest width between the horn bases, where the kouprey is exceeded by 192% in the bantin and 242% in the gaur. The bantin horns have turned-in tips like the kouprey but they are smaller than those of the gaur as well as of the kouprey. This is particularly evident in the circumference, where the gaur exceeds the bantin by 18.9% at the base and 21.2% near the tip. The outer ends of the gaur and bantin horn-cores are directed slightly posteriorly while those of the kouprey are directed slightly anteriorly.

Mandible.

The differences between the mandibles of the gaur and bantin are principally in the more hook-like shape of the coronoid in the gaur and its greater extension beyond the condyle. The angle of the ramus has not the bulging curve in the gaur that it has in the bantin. Unfortunately the mandible of the kouprey is missing. It must have been misplaced at the time the skeleton was packed up in the field. It is hoped at some future time to publish a supplementary note comparing the mandible of the kouprey with those of the gaur and the bantin.

COMPARATIVE MORPHOLOGY

The skull measurements have shown certain differences between the adult male kouprey, bantin, and gaur that we are comparing. Certain other differences, particularly differences in shape can best be described and in some cases shown most clearly by scale drawings.

The horns will be discussed and skulls of kouprey M.C.Z. 38108, bantin M.C.Z. 36669, and gaur M.C.Z. 36670 will be compared from the lateral, dorsal, ventral, and nuchal views.

Horns

The horns of the kouprey, the color and measurements of which have been referred to elsewhere, are distinctive on account of their shape and curve, as well as the corrugations on their lower part and the bushy frill of shredded horn near their tips. The horns leave the frontal at a reverse angle such that if their axis were prolonged they would very nearly intersect at right angles, as indicated by the dotted line in the ventral view. The horn curve is backward, outward, upward, forward, and finally upward, inward and slightly backward. In the gaur and bantin the curve is much simpler. In the gaur it is downward and then upward and backward and slightly inward near the tip. In the bantin it is downward then upward, slightly backward and definitely inward at the tip.

Gaur and bantin horns have a rough corrugated area close to their base but this is more extensive in the kouprey on the upper side of whose horn there are seven cross-ridges extending over an area of 170 mm. out from the base. The gaur shows two such ridges and traces of a third but all within 70 mm. of the base of the horn. The roughness of the bantin horns covers a similar area but does not form definite ridges. The cross-section of the kouprey horn is a flat oval at 100 mm. from the base, and circular at 100 mm. from the tip (see fig. 3). The cross-section of the gaur horns would be very similar and those of the bantin a rounded oval at the base. Our kouprey horns maintain their almost triangular flat oval cross-section up to within 300 mm. of the tips when they gradually change into the rounded cross-section. In the bantin the cross-section of the horn is almost

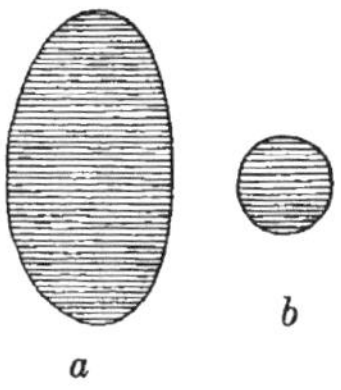

Fig. 3. Cross section of Kouprey horn (¼ nat. size); *a*. 100 mm. from base; *b*. 100 mm. from tip (M.C.Z. 38108).

circular 100 mm. from the base and the round form continues to the outer tip. In the gaur the flattened oval cross-section develops into the rounded form almost as rapidly as in the bantin so that the cross-section is virtually circular 200 mm. from the base and the rest of the way out to the tip. The sharp-pointed tips of the horns of the gaur and bantin are slenderer and taper less rapidly than those of the kouprey.

The hornsheath of the frontlet has three transverse ridges on each horn close to the base where the surface is rough and blackish. Beyond this the horn is smooth, glossy, and of a deep brownish yellow-green color with long slender black tips. The underside of the horn is flattened for the first 300 mm. from the base. The horn is ovate for 100 mm., and rounded for the last 300 mm. to the tip. The horn curve is similar in both koupreys.

The kouprey horn-core is flattened ventrally and has a number of heavy ridges or grooves extending out along its surface. The most striking feature of the horn-core is the almost knife-like primary keel that extends along its posterior edge for a distance of 355 mm. This is 9 mm. wide 100 mm. from the base and is best shown in cross-sections of the horn made at approximately 100 mm. intervals (Plate III fig. 2 and Plate IV). There is a slight anticlockwise torsion to the horn-core and there is a prominent ridge and groove that extend out for 280 mm. just 30 mm. above the line of the posterior keel. The horn sheath was held on very tightly on account of these keels and ridges set in grooves of the horn sheath and the tip of the horn-core reached a point in the sheath about 160 mm. from the tip just about in line with the place where the external fraying was most marked. This was clearly shown in an x-ray taken by Dr. S. C. Sossman of the Peter Bent Brigham Hospital of the frontlet loaned by the American Museum of Natural History.

In our comparison of the kouprey horns the frontlet obviously belonged to a younger animal than our specimen, and the horn sheaths left the skull at more nearly an angle of 125° than 105° of the older specimen.

The horn-core of this frontlet is broad and flattened. It has a slight anticlockwise torsion, the tip points slightly forward, has a number of deep grooves and shows definitely a small sharpened posterior ridge which marks an early stage of the development of the primary posterior keel so prominent in the older animal. The horn-core of the bantin is more rounded than that of the kouprey and has a more slender extremity with the tip pointing inward and back. There are a number of heavy grooves on the posterior and anterior sides of the horn-core. The bantin horn-core has a considerable anticlockwise torsion. The horn-core of the gaur is in general lozenge-shaped more similar to that of the kouprey with a slight anticlockwise torsion. It is in some respects exactly opposite to that of the kouprey in the fact that in cross-section it is rounded behind and more flattened anteriorly. It also has heavy grooves especially on the lower posterior and anterior surfaces. The curve of the gaur horn-core is upward and slightly backward toward the tip.

The horn-core of an old male wild yak (M. C. Z. 21578) has a number of deep grooves and many minor ones extending in roughly paralled lines in the long axis on all faces of the horn surface. The horn core tip is pointed slightly posteriorly with reference to the axis of the horn. The cross-section of the horn core base is a distinctly more rounded oval than that of the kouprey. The torsion of the horn-core is slightly counter clockwise.

The fraying of the horns near the tips has been mentioned in all descriptions of the kouprey and can be seen in the photographs of the mounted head (Urbain, 1937) as well as in our specimen, and the frontlet loaned us by the American Museum. It seems to be characteristic of the bulls although we know nothing as yet about the actual horn conditions in the cow kouprey. The fraying is undoubtedly caused by some habit of the kouprey which involves digging with or scraping the tips of the horns. The outer surface of the horn is shredded into fibers about 2 mm. wide in our specimen and these form an irregular collar completely around each of the horns. The horn has been reduced by almost 10 mm. of horn all the way round above the collar as a result of the habit that causes this condition. The American Museum frontlet (see Plate III, fig. 3 and Plate V) shows a similar fraying, but the horn has split off in much coarser strips than in the older bull, the individual shavings being 5 or 6 mm. wide. In both pairs of horns one side is more worn than the other; in the older bull it is the right side and in the younger one the left.

It would be interesting to find out what habit causes this fraying, whether it is associated with digging for special roots or whether it is limited to bulls and caused by some special behavior during the rutting season. It is a curious fact that extensive fraying of the horns is not to my knowledge associated with any other of the living Bovidae, not even with one of the wide range of domesticated races of cattle, nor is it shown in representations of cattle in art. The nearest approach to it is probably seen in an occasional very old bison where the tip of the horn is worn thin leaving a ridge of thicker horn a little below the tip.

In an effort to determine whether the kouprey horn sheath was different in hardness, structure, or chemical constituents from that of the gaur, we excised a small section of the outer part of the horn sheath from a similar place in the kouprey and the gaur. Thin sections were prepared for comparison under the microscope. The examination indicated no structural difference other than a greater softness in the kouprey horn which made it more easily scratched.

The two horns were tested for scratch hardness with the result that the kouprey proved to be 3½ on the mineralogist's Mohs' scale as compared with 4 for the gaur. Calcite has a hardness of 3 and fluorite of 4. The horns were tested for penetration hardness with a Rockwell B 1/8-inch ball resulting in 76 as the average reading for the kouprey and 82 for the gaur. In this test the higher reading indicates greater hardness.

The chemical analysis of the keratin in these two horns was not carried beyond that of obtaining a total nitrogen Kjeldahl analysis. The average value for the gaur was 15.21% and for the kouprey 14.93%. This difference of 0.3% is almost within the limits of experimental error. It would be of interest to have comparative analyses made for sulphur, etc., but such tests would have to be repeated many times and were not practical at this time.

Lateral View

In outline the distinctly depressed upper part of the frontal so characteristic of the gaur profile is found to a less degree in the bantin and is not evident in the kouprey. The frontal of the kouprey is flat and gradually slightly depressed at a point just above the orbits. When the skull without the mandible is horizontal, resting on a flat surface, the frontal part of the gaur and bantin skulls is heavier and higher from the horizontal base than that of the kouprey. The greatest height of the temporal fossa measured between the zygomatic process of the temporal and the frontal is 50 mm. in the bantin and gaur, and 31 mm. in the kouprey. The height of the frontal at this same point is 60 mm. in the gaur, 50 mm. in the bantin, and 40 mm. in the kouprey.

The horns and horn-cores are more anteriorly directed in the kouprey than in the bantin and gaur. The zygomatic process of the temporal in the bantin is light and sharply curved down anteriorly. In the gaur the same bone is thick and wide with a curve similar to that of the kouprey. The temporal crest is pronounced in the kouprey and slight in the gaur and bantin. The horns of the bantin and gaur are depressed after leaving the skull so that from a lateral view the auditory meatus is just visible while in the kouprey one can see the top suture of the parietal and temporal bone as in domestic cattle skulls.

The lachrymal is narrower as it approaches the orbit in the kouprey than in bantin or gaur. The narrowest external widths are respectively 22 mm., 32 mm., and 27 mm. in the gaur. The suture between the lachrymal and the maxilla, and the malar and the maxilla is very irregular in the bantin and gaur and makes in general a double curve with the high point at the malar-lachrymal junction. In the kouprey this suture is more regular and practically forms a continuous transverse line (Plate VI). The short maxillary tooth row of the kouprey has already been mentioned. In lateral view the nasals project 72 mm. in the kouprey as compared with 40 mm. in the bantin, and 60 mm. in the gaur. The premaxilla projects further forward in the kouprey than in the gaur or bantin and its ascending process extends up to the nasals where it has a contact for 10 mm. with the nasal. In the gaur and bantin the highest point of the premaxilla is 10 mm. below the nasal from which it is separated by the maxilla. The ascending process of the premaxilla of the bantin narrows toward the upper part, that of the gaur maintains its width and terminates in a squared end; in the kouprey it is more nearly like that of the gaur only wider and heavier with a differently shaped top as shown in the lateral view (Plate VI). In the kouprey and bantin if we erected a perpendicular at the labial alveolar midpoint of PM^2 it would intersect the infra-orbital foramen which is on the same horizontal plane as the premaxilla. In the gaur the infraorbital foramen is slightly more posterior and a perpendicular erected there would intersect the maxillary tooth row between the PM^2 and PM^3. In the bantin the paroccipital process is slender and pointed, in the gaur it is heavier with a flattened point, in the kouprey it is heaviest with a thickened anterior ridge.

Dorsal View

In the kouprey one is immediately struck by the long slender skull suggesting in outline that of Bos primigenius, Bos longifrons, and Bos indicus rather than the bison-like wide-frontal type which is characteristic of the gaur and bantin. The orbital sockets and the facial tuberosities are prominent (Plate VII).

There is no marked frontal eminence, and as already shown in the measurements, the posterior bases of the horns are near together. The figure shows clearly the angle at which the horn-cores branch off from the frontal and the absence of the prominent intercornual ridge so characteristic of the gaur and the bantin. There is a marked supraorbital groove, but this is most evident in the bantin and more marked in the gaur than in the kouprey. It appears to become more prominent with the age of the animal.

There is a slight depression of the frontal in the midline. This depression extends from the point on the median line that would be intersected by a line joining the supraorbital foramina forward to include the upper tips of the nasals and terminates at the anteriormost point of the frontal. There is no such depression in the bantin and a slight indication of one in the gaur.

The median frontal suture is fused, forming a raised ridge extending 90 mm. back from the point where this suture would be intersected by a line joining the supraorbital foramina. The extent of this fused ridge may be determined by the age of the specimen. No such prominent ridge is evident in the gaur or the bantin.

A horizontal line connecting the points on the orbital rims where the lachrymal and frontal articulate will pass very close to the upper tip of the nasals in the old adult kouprey and the adult bantin, whereas in the adult gaur the nasals extend 20 mm. above this line.

The angle at the upper end of the nasals where they articulate with the frontal is wider and less sharp in the kouprey and resembles more that in the yak than it does the gaur and bantin (see fig. 4). The kouprey nasals are conspicuous on account of their great length. They are somewhat arched from side to side. Where the frontals, nasals and upper part of the lachrymal come together there is an ethmoid vacuity, an open triangle about 15 mm. on each side (Plate VII.). The spines at the lower end of the kouprey nasals being long in the center represent a primitive condition and resemble most closely the condition in the yak. The gaur has notched anterior nasal spines longest in the center, but the bantin and buffalo have four anterior nasal spines with the longer spines outside and shorter ones in the center (see fig. 4).

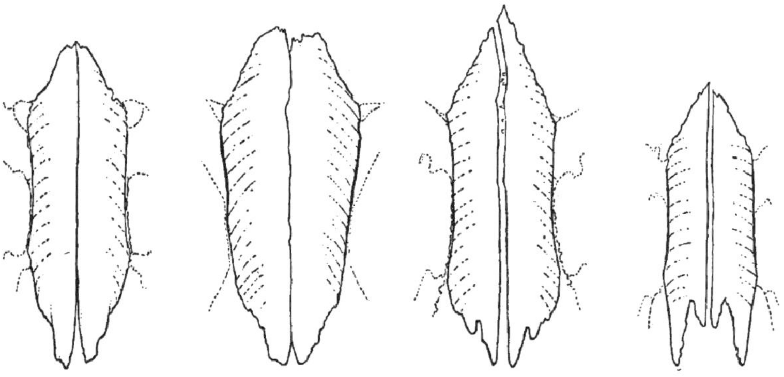

Text Fig. 4. (lengthwise)

Fig. 4. Nasals (⅙ nat. size) of *a*. Kouprey (M.C.Z. 38108); *b*. Wild Yak (M.C.Z. 21578); *c*. Gaur (M.C.Z. 36670); *d*. Bantin (M.C.Z. 36669).

At the orbital margin of the lachrymal there are rough tuberosities which form a distinct notch in the mid-lachrymal of both the gaur and the bantin. There are smaller notches in the orbital rim at the point where the lachrymal articulates with the frontal and the malar bone. In the kouprey there is a large notch on the orbital rim at the intersection of the lachrymal and malar. There is a small notch where the lachrymal and frontal articulate, and there is no further notch in the orbital part of the lachrymal.

The articulation of the nasal process of the premaxilla with the nasals has been mentioned in discussing the lateral view. The body of the premaxilla is heavier in the gaur and the kouprey than in the bantin. In the kouprey the anterior tip of the premaxilla extends 70 mm. beyond the tip of the nasals, in the bantin 96 mm., in the gaur 85 mm. The incisive fissure in the kouprey, yak and bantin is more bottle-shaped than in the gaur where it is narrower with the sides more nearly parallel. The palatine fissure is closely similar in size and length in the kouprey and gaur. In the bantin it is 10 mm. shorter. The palatine process is discussed in the description of the ventral view. The lower end of the body of the premaxilla is more nearly square in the kouprey, yak and gaur than in the bantin where it is more rounded.

Basal View

As in the dorsal view one is immediately struck by the long slender outline of the kouprey skull. Also the backward angle at which the horns leave the frontal is very evident. (Plates VIII, IX and X). The heavy ridging of the paroccipital process in the kouprey referred to in discussion of the lateral view is clearly shown in the ventral view and contrasts with a more slender process in the figures of the gaur and bantin.

The basilar tubercles are more prominent in the gaur than in the bantin and kouprey. In the kouprey the bulla ossea is laterally compressed with a sharp ventral ridge and pointed anteriorly with a prominent spine (Van der Klaauw, 1931, p. 63; Bondy, 1907, p. 381). In the gaur they are considerably inflated without the ridge and with a less prominent spine, while in the bantin they are even more inflated and have a smoother surface than in the gaur. The wild yak (Acad. Nat. Sci. 17311) has laterally compressed bullae more closely resembling those of the kouprey.

The narrowing of the frontal between the horn-cores and the orbits is more marked in the bantin and the kouprey than in the gaur. This is shown by the visible space between the frontal and the zygomatic arch in the comparative basal views.

In the kouprey the crests formed by the pterygoid and terminal processes of the palatine bones are more widely separated posteriorly than they are in the gaur and bantin where these crests are more nearly parallel. Also in the kouprey the center of the horizontal part of the palatine bone extends between the terminal crests of the palatine bone over the posterior nares for 33 mm. back of a horizontal line crossing the palate posterior to the third molars. In the gaur and bantin a similar measurement made in the palatal midline is 10 mm.

In kouprey, gaur and bantin there is a small vomer well separated from the palate.

The lachrymal bullae are broken in the gaur skull that we are using for comparison, but in other gaur skulls these bullae appear to be more prominent from the basal view than they are in the kouprey and bantin.

The kouprey has a prominent maxillary tuberosity. This consists of a flattened area the width of the tooth row that extends back for 14 mm. behind upper M^3. In the bantin and the gaur the maxilla slopes away almost directly behind the upper M^3. The age of an animal has an important bearing on this point.

The interior palatine foramina are located on the suture between the palatine process of the premaxilla and the horizontal part of the palatine bone in the kouprey, yak, and gaur but posterior to this suture in the bantin. This palatal suture in the kouprey would coincide very closely with a horizontal line across the palate between the first and second molars. In the gaur the palatal suture is curved in a general U shape anteriorly. At the middle of the palate the top of the U will almost intersect a horizontal line across the palate between the fourth premolars and the first molars. In the bantin the palatal suture is placed more anteriorly as in the gaur but the palatal outline of it is more nearly horizontal as in the kouprey than U shaped as in the gaur.

If a rigid base line is drawn to connect the midpoint of the palatal tip of the premaxilla with the most posterior point of the midline of the horizontal part of the palatal bone, in the kouprey such a line would only touch the skull at the two end points. At the narrowing of the palate above the premaxilla the palatal surface would be depressed 25 mm. from such a base line. In the gaur such a rigid palatal base line would be in contact with the palate for almost half its length posteriorly and the most depressed point close to the premaxilla would be 13 mm. below the base line. In the bantin the portion of the palate between the tooth rows is even flatter than in the gaur and a base line cannot connect the end points of the palate between the first premolars. The depression in the bantin palate at its narrow point is more localized even than in the gaur and about 12 mm. below the palatal base line.

The kouprey palate not only has a marked slope forward as far as the point of intersection of the palatine process of the premaxilla but it is slightly more curved horizontally than the wider and more flattened palates of the gaur and bantin.

The short maxillary tooth row of the kouprey is very evident in the ventral view. The tooth row in the bantin is intermediate between that of the kouprey and the long one of the gaur. The proportionately long premaxilla in the kouprey is clearly evident in the palatal view. The palatine processes of the premaxilla are slender in the kouprey and bantin and lighter than those in the gaur. The articulation of these (Plates VIII, IX and X) with the maxilla is quite different in the kouprey and bantin from that of the gaur. The distance between the anterior surface of the first premolar and the articulation of the palatal part of the body of the premaxilla with the palatine process of the maxilla is 65 mm. in the gaur, 68 mm. in the kouprey, and in the bantin, which is asymmetrical in this region, 83 mm. on the left side and 73 mm. on the right side. The suture line of articulation is more horizontal in the kouprey and the gaur and on an angle for the complete palatal width of the bones in the bantin as can be clearly seen in the plates. The palatine fissure, the lower end of the body of the premaxilla, and the incisive fissure have been mentioned in the discussion of the dorsal views.

The kouprey frontlet from The American Museum was cut off of the skull in such a way as to reveal clearly the top of the brain case (Pl. V, fig. 1). From this we can see that the parietal covers between a third and a half of the top of the brain. Its frontal suture is shown by a dotted line. The heavy sinuses in the occiput can be seen in this section as well as the upper side of the temporal fossae which open on to the occiput at points comparatively close together.

Nuchal View

The differences between the kouprey, gaur, and bantin can be readily seen by comparison (Plate XI). In each case the occiput consists of two distinct parts separated by a narrowing at the point where the temporal fossae open freely on to the occiput. This significant narrowing is greatest in the kouprey and greater in the gaur than in the bantin. Above the narrowing lies the parieto-frontal part of the occiput which in the kouprey is very much restricted by the horn sheaths coming so close together that the visible part of the bone, except for the edges under the horn-casings, almost makes a square with a parietal surface smooth and slightly rounded. By contrast in the gaur and bantin the large parieto-frontal part of the occiput is long and continuous with the horn-cores connected by a prominent intercornual bony ridge. (In some bantins there is an extension of horn sheath across this area.) In the gaur and bantin the sutures between the occiput and the frontal are clearly visible, and the occiput extends slightly higher in the bantin than in the gaur.

The lower part is made up entirely of the occipital bone, and as already mentioned is separated from the upper by two deep notches leading from the temporal fossae. The lower part is bounded above by the ridges of the occipital crest, which in the kouprey are prominent and subtriangular with sloping sides like a steep building roof. In the bantin this occipital crest is flattened, in no way prominent, and almost semi-circular in shape. In the gaur the side crests are more marked and the condition is somewhat intermediate between that of the bantin and that of the kouprey. In the kouprey there is a marked external occipital protuberance with a slender but definite median occipital crest extending down to the foramen magnum. In the gaur and bantin the external occipital protuberance is not marked and there are slight traces of a median occipital crest more evident in the gaur than in the bantin.

Both the supra-cristal parts of the occiput and the infra-cristal part are wider and more extensive in the gaur and the bantin than in the kouprey. The occipital condyles of the kouprey are more triangular in the nuchal view than are those of the gaur or bantin. In size they are very similar to those of the bantin. The foramen magnum is proportionately narrower in the kouprey than in the bantin or gaur as already shown by the measurements. The anterior edge of the foramen magnum is almost semi-circular for its full width in the gaur whereas in the bantin the semi-circular opening is only 12 mm. in the center of the basilar part of the occipital bone. In the kouprey the shape of the anterior edge of the foramen magnum is intermediate between the forms of the gaur and bantin. The paraoccipital process is heavily ridged in the kouprey and squared on the ends. It is slenderer in the gaur and even more slender in the bantin without squaring on the ends.

DENTITION

The maxillary tooth row in the kouprey is markedly shorter than that of the bantin or gaur. Its length is 118 mm. as compared with 134 and 156 in the other two forms. The slight general curve of the maxillary tooth row is similar in the kouprey, bantin and gaur. In all three the molars are nearly parallel, with the M^3's slightly closer to one another than M^1 and M^2. PM^3 and PM^4 are almost parallel continuing the line of M^1 and M^2, but PM^2 is turned inward on the lingual side bringing the PM^2 on each side closer together than any other teeth on opposite sides of the palate. In the gaur and bantin the masticatory surface slopes downward and inward so that the labial edge is prominent and sharp. There is less slope in the kouprey because the teeth are more worn on account of the age of the specimen.

TABLE V

COMPARATIVE TOOTH MEASUREMENTS

	Kouprey (M.C.Z. 38108)		Bantin (M.C.Z. 36669)		Gaur (M.C.Z. 36670)		Absolute difference between (A.P.) measurements of	
	Antero-posterior	Labio-lingual	Antero-posterior	Labio-lingual	Antero-posterior	Labio-lingual	Bantin and Kouprey	Gaur and Kouprey
1. Tooth widths								
PM ^{2}	17		18		21		+1	+4
PM ^{3}	15.5		19.5		19.5		+4	+4
PM ^{4}	16		18		21.1		+2	+5.1
M ^{1}	20	22	23	21	27	26	+3	+7
M ^{2}	23	23	27.5	20	31.5	26	+4.5	+8.5
M ^{3}	25.5	23	26	19	36	26	+0.5	+10.5
2. Length of combined molars	70.5		76		94.5			
3. Length of combined premolars	50		58		61.5			

The length of the combined molars in the kouprey is 70.5 mm. as compared with 76 in the bantin and 94.5 in the gaur. In the length of the combined molars, while 5.5 mm. is a significant difference between the kouprey and the bantin, the most striking thing is the difference between these two and the gaur in this measurement.

The length of the combined premolars in the kouprey is 50 mm. as compared with 58 mm. in the bantin and 61.5 in the gaur. In the premolars the gaur and bantin are within 3.5 mm. of each other while the kouprey is 8 mm. shorter than the bantin. In the kouprey upper PM^2 and PM^3 are very much worn as is M^1. PM^4 is very much crowded and overlaps M^1 on the contact side in each tooth row. Upper PM^2 on the left side is missing. The socket shows two root spaces, a single one forward and a wider one possibly for fused roots nearest to PM^3. The tooth wear of the kouprey shows that we are dealing with an older animal than the adult gaur and bantin with which we are making the comparison.

A comparison of the individual tooth widths in the three forms shows each of the PM's in the kouprey to be narrower than in the bantin, and PM^2 and PM^4 in the bantin to be narrower than in the gaur. In the kouprey and the gaur PM^3 is narrower than PM^2. In the gaur PM^3 is to a similar degree narrower than PM^4

but in the kouprey PM^3 is only slightly narrower than PM^4. In the bantin, however, PM^3 is wider than either PM^2 or PM^4. In the gaur and bantin PM^2 and PM^4 are the same width. In the kouprey PM^4 is clearly narrower than PM^2 and as would be expected shows considerably less wear. If we measure the PM's labio-lingually we find PM^2 and PM^3 similar in the bantin and kouprey but flatter than in the gaur. PM^4 on the other hand is 3 mm. flatter in the bantin than in the kouprey but the kouprey is still 3 mm. flatter than the gaur. The accompanying figure shows the difference in crown pattern in PM^4 of each of the three oxen that we are comparing.

The infundibulum (see fig. 5d.) is in general horse-shoe-shaped in the kouprey, and more flattened, of an irregular elongated shape in bantin and gaur. The enamel ridges on the top of the labial face of PM^4 in the kouprey are sharply infolded in two places. In the gaur this surface is folded in two prominent deep U curves. A similar condition occurs in the bantin but the valleys of the U's are shallower. The lingual surface of PM^4 in the kouprey is more pointedly arched, whereas it is quite flattened in the gaur and bantin.

To sum up some differences in the premolars: The kouprey has narrower (antero-posteriorly) ones than the gaur and bantin and also differs in having PM^4 narrower than PM^2. Our comparison also shows that PM^3 in the bantin is greater in width than PM^2 and PM^4 in the bantin whereas it is smaller in the gaur and kouprey. The crown pattern of PM^4 in the kouprey is simpler and more primitive than in the two others, and the tooth is greatly flattened, being almost rectangular with its long axis labio-lingual.

In comparing the shapes of the individual upper molars M^1 in the kouprey is 20 mm. wide antero-posteriorly and 22 mm. wide labio-lingually. This makes it slightly rectangular (l.l.). In the bantin $M^1 = 23$ (a.p.) 21 (l.l.) which makes it slightly rectangular (a.p.). In the gaur $M^1 = 27$ (a.p.) 26 (l.l.), which makes it like the bantin slightly rectangular (a.p.).

M^2 in the kouprey measures 23 mm. (a.p.) 23 mm. (l.l.), which makes it square. In the bantin $M^2 = 27.5$ (a.p.) 20 (l.l.) which makes a strong rectangular shape (a.p.). In the gaur $M^2 = 31.5$ (a.p.) 26 (l.l.) which makes a definite rectangular shape (a.p.) almost as marked as in the bantin.

M^3 in the kouprey measures 25.5 (a.p.) 23 mm. (l.l.) which makes it slightly rectangular (a.p.). In the bantin $M^3 = 26$ (a.p.) 19 (l.l.) which makes a strong rectangular shape (a.p.) similar to M^2. In the gaur $M^3 = 36$ (a.p.) 26 (l.l.) which makes an elongated rectangular shape (a.p.).

In directly comparing the individual upper molar widths (a.p.) in the three forms, each of those in the kouprey is narrower than in the bantin and each of those in the bantin narrower than in the gaur, except that the width (a.p.) of M^3 is closely similar in the kouprey and bantin. In comparing the width (a.p.) of the three molars in the tooth row of each individual form we find that in the kouprey and gaur M^2 is intermediate in width (a.p.) between M^1 and M^3. This is not the case in the bantin where M^2 is wider than M^3 as well as M^1. If the gaur-kouprey pattern was followed in the bantin one would expect M^3 in the bantin to have a width (a.p.) of 31.5 but instead it has a width of 26 mm. which is less than M^2.

The labio-lingual widths show all the molars in the kouprey to be nearly square whereas this is only true of M^1 in the gaur and bantin.

If we compare the masticatory surface or crown pattern of the molars in our three forms we must allow for a difference in wear which is greatest in the kouprey. The bantin and gaur show about the same degree of wear. (See figs. 6 d, e, f.).

The pattern of M^1 in the kouprey is destroyed by the extreme wear. The labio-lingually rectangular shape of this tooth has already been referred to. The pattern of M^1 in the gaur and bantin is surprisingly similar. The accessory column is more prominent in the gaur as is the lingual fold of the posterior infundibulum which is smaller and less prominent in the bantin.

The pattern of M^2 in the kouprey has simplified horse shoe-shaped infundibula with a very small comma-shaped accessory column that does extend down to the roots of the tooth. In the gaur and bantin the infundibula are of very similar pattern, wider on the lingual than on the contact sides of the tooth. The accessory column is especially prominent with the shape of a large comma in the gaur and extends down to the roots. In the bantin the shape is more that of a triangle with points drawn out. It is more prominent than that of the kouprey in the crown pattern and extends down to the roots in a similar manner.

The shape of M^3 is definitely more nearly square in the kouprey than in the bantin or gaur. The crown pattern in the kouprey shows a small accessory column less prominent than that in the bantin and much less than that in the gaur. The almost triangular surface of this column is quite similar in the kouprey and the bantin where it is comma-shaped in the gaur. The anterior infundibulum in M^3 of the kouprey is again horseshoe-shaped or yak-like much as in M^2. The posterior infundibulum has a marked extra fold on the posterior lingual side which accounts for the shape being so different from the anterior one. In the gaur M^3 the posterior infundibulum shows the beginnings of the extra fold but it is in no way as distinct as in the kouprey. Both infundibula of the gaur are wider and more irregular in external shape than in the kouprey. The anterior infundibulum is quite similar in the M^3 of the gaur and the bantin except that in the gaur there is a deeper U-shaped curve on the labial side than in the bantin, and the bantin has a small notch on the lingual side. The bantin likewise has a small notch in the anterior side of its posterior infundibulum and very little hollowing on the labial side. The posterior lingual part has a small trace of the extra fold already mentioned as evident in the gaur and conspicuous in the kouprey.

If we compare the PM^4 crown pattern (see fig. 5) of the kouprey with others that are figured, the shape of the infundibulum and proportions of the

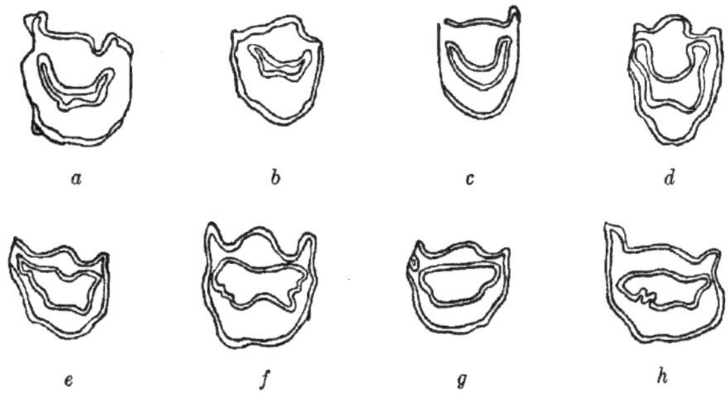

Fig. 5. Crown patterns (nat. size) of left upper PM^4 of 3 fossil and 4 living bovine genera.

a. Hemibos (A.M.N.H. 19963)
b. Leptobos (A.M.N.H. 19816)
c. Proleptobos birmanicus (after Pilgrim, Br. Mus. No. M10909)
d. Novibos sauveli (M.C.Z. 38108)
e. Bibos banteng, sub. sp.? (M.C.Z. 36669)
f. Bibos gaurus readi (M.C.Z. 36670)
g. Bos taurus (M.C.Z. 8388)
h. Poephagus grunniens mutus (Acad. Nat. Nat. Sci. 17311)

flattened tooth are most like that of Proleptobos. The enamel fold on the labial side is most like a flattened Bibos.

The crown pattern of M^1 in the kouprey is too worn for comparison but its external shape is most clearly approached by that of Proleptobos and to a less degree the bantin, zebu and the domesticated yak.

The crown pattern of M^2 in the kouprey is such that the enamel folds on the labial side with a prominent median notch suggest most (fig. 6 c) Proleptobos, the infundibulum and in general the shape of the lingual side suggest that of the domesticated yak, although the kouprey's M^2 does not seem to fit into any of the indicated patterns as closely as that of Proleptobos.

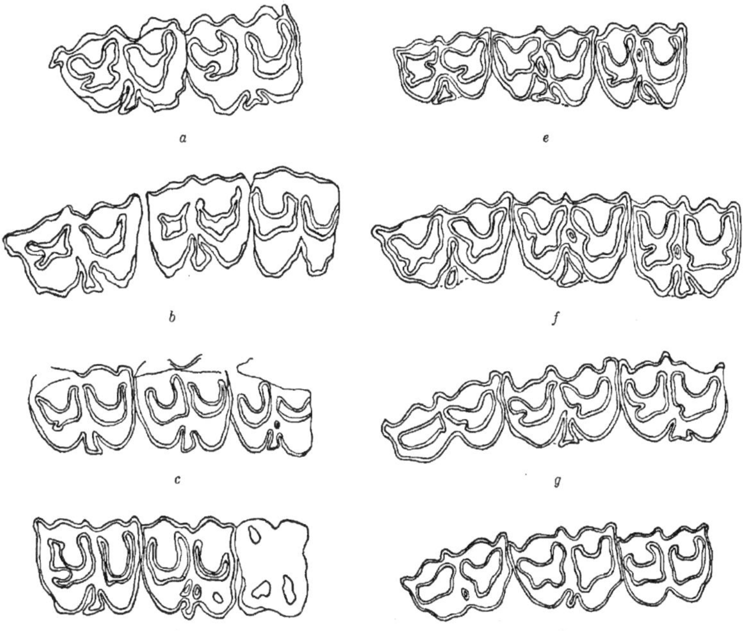

Fig. 6. Crown patterns (nat. size) of left upper molars of 3 fossil and 3 living bovine genera.

a. Hemibos acuticornis (A.M.N.H. 19963)
b. Leptobos falconeri (A.M.N.H. 19816)
c. Proleptobos birmanicus (after Pilgrim) (Br. Mus. No. M10909)
d. Novibos sauveli (M.C.Z. 38108)
e. Bibos banteng subsp. ? (M.C.Z. 36669)
f. Bibos gaurus readi (M.C.Z. 36670)
g. Poephagus grunniens mutus (Phil. Acad. Nat. Sci. 17311)
h. Poephagus grunniens (A.M.N.H. 10730)

The crown pattern of M^3 in the kouprey most nearly corresponds with Bos indicus in the enamel folding of the labial side and in the shape of the posterior infundibulum. The shape of the lingual surface more nearly resembles that of Bubalus.

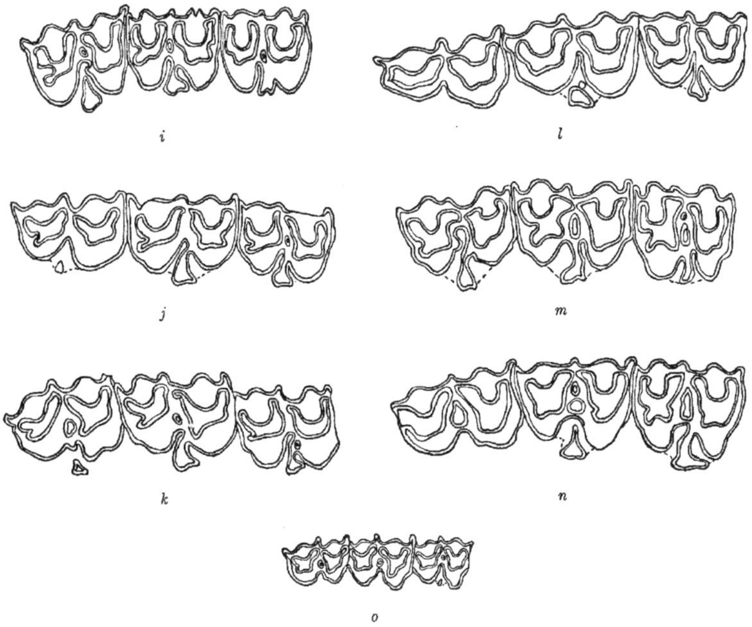

Fig. 7. Crown patterns (nat. size) of left upper molars of five living bovine genera. See p. 259.

i. Bos indicus (M.C.Z. 8387)
j. Bos taurus (M.C.Z. 8388)
k. Bos taurus (M.C.Z. 3)
l. Bison bison (M.C.Z. 11)
m. Bubalus bubalis (M.C.Z. 38768)
n. Syncerus aequinoctialis (M.C.Z. 11611)
o. Anoa quarlesi (M.C.Z. 27999)

SKELETON

Vertebrae

In the kouprey the 7th cervical vertebra has a spinous process 142 mm. in greatest length from the vertebral foramen. The spine is slightly backwardly directed and the vertebra differs in no essential way from the 7th cervical in the domestic ox or from Bos primigenius as figured by Reynolds (1939, p. 19, fig. 70). The spines of 1st, 2nd, 3rd, and 4th thoracic (dorsal) vertebrae are very long, the greatest length of the neural spine measured from the vertebral foramen, being 320 mm., 352 mm., 355 mm., and 350 mm., respectively. The sixth lumbar has a neural spine of 85 mm. in length and the transverse process is 130 mm. long measured from the foramen magnum. The 6th lumbar vertebra is essentially similar to that of the domestic ox. The sacrum will be discussed with the pelvis.

Scapula

The scapula of the kouprey is narrower and not so broadly triangular as that of the gaur and bantin (see fig. 8). The spine is straighter and less folded, and the supraspinous fossa narrower than in the bantin and gaur. The length of the neck measured from the coracoid edge of the glenoid cavity to the base of the spine below the acromion is 67 mm. in the kouprey, 12 mm. less than in the bantin and 13 mm. less than in the gaur. The tuber scapulae is more prominent in the kouprey and extends out from the nearest edge of the glenoid cavity in the kouprey 4 mm. more than in the bantin and 7 mm. more than in the gaur (see Table VI). The glenoid cavity of the kouprey is similar in shape to that of the gaur and differs from that of the bantin.

Metapodials

According to Schertz as interpreted by Reynolds, Bos is characterized by (a) a massive character of the metacarpal, (b) a relatively rectangular articulating surface for the carpus and one which does not taper away rapidly outwards, (c) the metacarpal is widest at the lower end. The kouprey metacarpal is proportionately long and more slender than most of the 18 Bison and Bos metacarpals figured by Reynolds. The articulating surface of the carpus shows the tapering characteristic of the bison. The fact that the bone is very nearly as wide at the epiphysial junction as at the lower end is another bison-like characteristic.

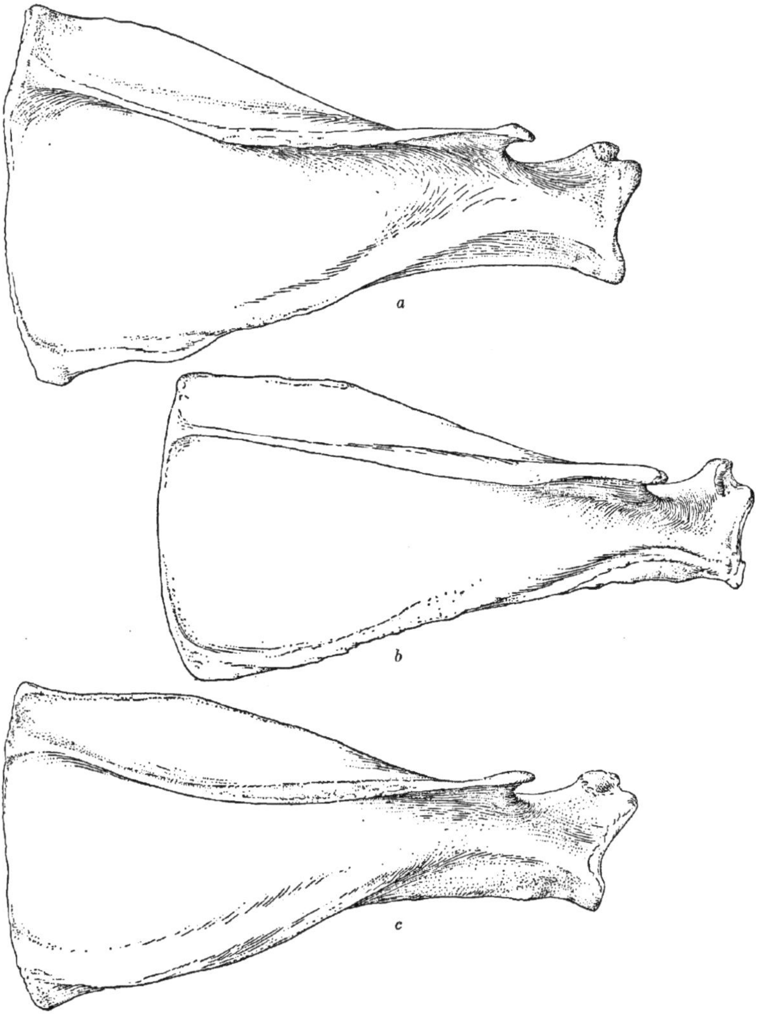

Fig. 8. See p. 462. Right scapulas (⅓ nat. size) of: *a*. Gaur (M.C.Z. 36670); *b*. Kouprey (M.C.Z. 38108); *c*. Bantin (M.C.Z. 36669).

The kouprey metacarpal corresponds in general proportions with No. M. 12105 right (labelled Bison), Mitcham, and figured (1939, Fig. 16, No. 13) by Reynolds. Its measurements are: 212 mm. length, 68 upper end, 37 middle of shaft, 65 at suture, 65.5 at distal end. This specimen was found associated with the slender metatarsal (B.M. 4090) mentioned below in the discussion of the metatarsals. As can be seen by the measurements, the kouprey metacarpal is considerably slenderer than that of the gaur (Am. Mus. Nat. Hist. 113747) with which it has been compared. The lesser width at the distal end would make it more like Bison than Bos according to Schertz and Reynolds. The articulating surface of the gaur metacarpal (Am. Mus. Nat. Hist. 113747) is relatively more rectangular than that of the kouprey.

The metatarsals of the kouprey have a prominent groove for the extensor tendon which extends along the whole shaft, and while most conspicuous at the lower end of the bone, it is of practically uniform depth. A similar groove not quite so prominent is found in the gaur and bantin metatarsal. The shaft is slenderer than in the gaur or bantin although more nearly like the latter. The distance from the nutrient foramen in the groove to the distal end between the condyles is proportionately greater in the gaur than in the bantin or kouprey.

The shaft of the kouprey metatarsal is slenderer than Bos and Bison metatarsals figured by Reynolds with the exception of one (B.M. 4090) from Mitcham, Surrey, which measures 33 mm. at the middle of the shaft and 61.5 at the distal end. The upper end is imperfect. This Mitcham specimen is attributed to "Bos longifrons."

To sum up, the metapodials of the kouprey show some bison-like features as interpreted by Schertz and Reynolds. They are more like those of the bantin and the gaur, and more slender than either one (for measurements see Table VII.)

The following small bones[1] of the fore and hind limb of the kouprey show no significant differences from the forms characteristic of ruminants: (2) radial carpals, (2) intermediate carpals, (1) ulnar carpal, (2) fused second and third carpals, (2) fourth carpals, (2) metacarpals, (4) first phalanges, (2) tibial tarsals, (2) fibular tarsals, (2) fused central and fourth tarsals, (2) fused second and third tarsals, (2) large metatarsals, and (4) first phalanges.

Humerus

The right humerus of the kouprey has the general form characteristic of ruminants as has the left radio-ulna. The humerus is smaller than that of the

[1]It is possible that in packing these small bones in the field, similar bones of an adult bantin might have gotten mixed with them.

gaur and bantin with which its measurements have been compared and the greater tuberosity in the gaur and bantin projects twice as far above the lesser tuberosity as it does in the kouprey.

Radio-Ulna

The olecranon of both gaur and bantin is longer than that of the kouprey after allowing for difference in size in the bones that are compared. On the anterior edge of the proximal end of the radius, there are three bony peaks or processes which differ in the radii we are comparing. In the kouprey the top of the lateral ligament tuberosity is 5 mm. higher than the coronoid process with a small notch between the two, and this tuberosity in turn is 5 mm. higher than the proximal part of the radial tuberosity.

In the bantin the coronoid tuberosity is highest with a big notch between it and the top of the lateral ligament tuberosity which is 4 mm. lower on the other side. The proximal part of the radial tuberosity is 5 mm. lower than the coronoid.

In the gaur the top of the lateral ligament tuberosity is the highest as in the kouprey. In the gaur there is a big notch between it and the coronoid process which is 2 mm. lower. Such a notch was also found well developed in the bantin. The proximal part of the radial tuberosity is 5 mm. lower than the lateral ligament tuberosity in gaur and bantin where it was 10 mm. less in the kouprey.

In the line of fusion between radius and ulna the kouprey and bantin are very similar to each other with proximal interosseous spaces of close to equal size but smaller in the kouprey. The gaur, although somewhat larger in bone as seen by the measurements, has an interosseous space more than three times the size of the two others.

Femur

The general morphology of the kouprey is similar to that of the gaur and bantin although there is considerable difference in the transverse measurements of the widths at the condyles. A weaker femur head with a smaller diameter is found in the kouprey. Although the kouprey's femur is smaller and lighter than the other two, a comparison of the measurements between the nutrient foramen and the lesser trochanter indicates a marked lengthening in this part of the shaft of the femur in both the gaur and the bantin as compared with the kouprey.

Tibia

The right tibia of the kouprey has the general form characteristic of ruminants. The tibia is smaller than that of the gaur and bantin with which its measurements have been compared. At the proximal end the nutrient foramen is 20 mm. away from the sulcus muscularis in both kouprey and bantin while it is only 6 mm. away in the gaur. At the distal end the articular surfaces of the lateral malleolus or the primitive distal end of the fibula are smaller and less flattened in the kouprey and bantin than in the gaur. The nutrient foramen on the anterior side of the distal end of the tibia is 10 mm. from the nearest articular groove in the kouprey and bantin and 25 mm. in the gaur. There is more of a process to indicate the head of the fibula in the kouprey and bantin than in the gaur.

On the basis of comparative measurements of the longbones (see Table VIII), those of the gaur are longest and stoutest with the exception of the tibia which is of equal length with that of the bantin. In every case the longbones of the kouprey are the shortest and slenderest of the three. In length of humerus the bantin is 40 mm. greater than the kouprey and the gaur 28 mm. greater than the bantin. There is a very similar difference in the length of femora. In length of radius the bantin exceeds the kouprey by 18 mm. and is only exceeded by 6 mm. in the gaur. In the length of ulna the bantin exceeds the kouprey by 34 mm. and is only exceeded by 8 mm. in the gaur. In length of tibia the bantin and gaur are similar and both exceed the kouprey by 50 mm.

In terms of the proportion of the humerus and femur in all three the tibia of the gaur is greatly shortened and the radius and ulna are somewhat shortened. The radius and ulna of the bantin are also shortened but proportionately less than the gaur. In the kouprey the tibia is shortened a little, while the radius has lengthened considerably and the ulna a little.

In the morphology of the humerus and femur, there are definite points in which the kouprey differs from the gaur and bantin while they resemble each other. The same is true of the proximal end of the radius. In the interosseous space between radius and ulna and in the position of two foramina on the tibia the kouprey and bantin are similar and differ from the gaur.

Pelvis and Sacrum

The pelvic aperture of the kouprey, bantin, and gaur is almost rectangular with the ilia almost parallel to each other. In the kouprey there is 40 mm. difference between the width and the length of the rectangle, in the gaur which is larger there is a 52 mm. difference, while in the bantin only a 21 mm. difference and in the Jersey bull there is only a 5 mm. difference. For measurements see Table IX.

The greatest transverse width of the pelvic surface of the ilium is least in the kouprey and greatest in the gaur. From the measurements of the pelvic surface of the wing of the sacrum and ilium we note that the wing of the sacrum is 85 mm. for the kouprey and 90 for the gaur whereas it is only 67 for the bantin and 70 for the Jersey bull. The measurement on the pelvic surface of the ilium from the outer point of the wing of the sacrum to the tuber coxae is 110 for the kouprey and 125 for the gaur while it is 135 for the bantin and 150 for the Jersey bull. This makes the difference between the widths of ilium and sacral wings 25 mm. in the kouprey, 35 mm. in the gaur, 68 mm. in the bantin and 80 mm. in the Jersey bull.

On the dorsal side of the wing of the ilium the truncated tuber sacrale is shorter in the bantin and kouprey by 25 mm. than it is in the gaur and 15 mm. shorter than it is in the Jersey bull.

The ventro-dorsal width of the ilium shaft at its narrowest point is 21 mm. wider in the gaur than in the kouprey and 13 mm. wider than in the bantin. In thickness, however, the gaur's ilium shaft is greatly flattened and is within 1 mm. less than that of the kouprey but 6 mm. less than the bantin and 10 mm. less than the Jersey bull.

The combined length of the five fused sacral vertebrae measured on the ventral side is 32 mm. less in the kouprey than in the bantin and 53 mm. less than in the gaur, whereas on the dorsal side the length of median crest formed by fused sacral spines makes the kouprey 12 mm. less than the bantin and 39 mm. less than the gaur.

The curve on the ventral side of the fused sacrals is slightly greater in the gaur than in the bantin and kouprey. The median crest with its thick and rough margin narrows down to a width of less than 5 mm. between the spines of the 4th and 5th sacral vertebrae in the kouprey and bantin, but not in the gaur or Jersey bull. There is also a dip in the curvature of the median crest in the bantin and kouprey at the narrow point just referred to. This is indicated in the Jersey bull, but not in the gaur.

The lateral crest formed by the fusion of the articular processes is discontinuous in the kouprey, gaur and bantin, but continuous in the Jersey bull.

The dorsal medial crest of the sacrum projects 80 mm. above the lateral crest in the kouprey, 75 mm. in the bantin, and 100 mm. in the gaur, but only 55 mm. in the Jersey bull.

There is a marked continuous ventral ridge on the pelvic symphysis of the gaur and Jersey bull, whereas for a distance of 25 mm. posterior to the prominent symphysis pubis there is only a very slight ventral ridge which becomes again more marked beyond that point in the kouprey and bantin. The obturator foramen is in the form of a full oval in the gaur and Jersey bull and a flattened oval in the kouprey and bantin.

Summary. The pelvis and sacrum of the kouprey are smaller than those of the gaur, bantin, and Jersey bull with which they have been compared. On the ventral side the proportions of the pelvic aperture and the wings of the sacrum and ilium are similar in the gaur and kouprey and differ from the bantin and Jersey bull, whereas the ridge on the pelvic symphysis and the shape of the obturator foramen are similar in the bantin and kouprey, differing from those of the gaur and Jersey bull. On the dorsal side the bantin and kouprey have a shorter tuber sacrale than the gaur or Jersey bull, also the length projection and shape of the dorsal median crest are closely similar in the kouprey and bantin but differ from the gaur and Jersey bull. The kouprey is distinct from the other three in its markedly shorter total ventral length of fused sacral vertebrae. The bantin is closer to it in this respect than the gaur.

TABLE VI

SCAPULA MEASUREMENTS

	Kouprey M.C.Z. 38108	Bantin M.C.Z. 36669	Gaur M.C.Z. 36670
61. Width between anterior and posterior angles	200	218	246
62. Width of supra-scapula	60	—	—
63. Length of neck to base of spine	67	79	80
64. Projection of tuber scapulae	17	13	10
65. Glenoid cavity long axis	78	70	79
66. Glenoid cavity transverse	60	64	60

TABLE VII

METAPODIALS

	Metacarpals		*Metatarsals*		
	Kouprey M.C.Z. 38108	Gaur A.M.N.H. 113747	Kouprey M.C.Z. 38108	Bantin A.M.N.H. 54470	Gaur A.M.N.H. 113147
67. Length	230	240	258	256	266
68. Width at upper end	66	78	56	60	66
69. Width at middle of shaft	42	48	34	37	40
70. Width at suture between shaft and lower epiphysis	61	76	60	61	73
71. Width at distal end	61.5	74	58	59.5	72.5
72. Nutrient foramen to distal end between condyles	29	40	28	32	39

TABLE VIII

MEASUREMENTS OF LONGBONES

	Kouprey M.C.Z. 38108	Bantin A.M.N.H. 54470	Gaur A.M.N.H. 113147
HUMERUS			
73. Length from great tuberosity to external condyle (anterior side)	320	360	388
74. Transverse measurement of trochlea	92	106	112
75. Transverse measurement of proximal end from head to great tuberosity	126	147	150
76. Projection of lateral (great) tuberosity over medial (lesser) tuberosity	20	40	47
RADIO–ULNA			
77. Length of radius measured along middle of anterior face	314	332	338
78. Maximum width of radius at anterior end	90	97	106
79. Width across distal end of radius and ulna	81.5	75	86
80. Length of ulna measured along curvature	416	450	458
81. From processus anconaeus to highest point of olecranon	106	129	128
82. Length of proximal interosseous space	17	23	70
FEMUR			
83. Length measured from head to internal condyle	368	406	438
84. Width at condyle	107	113	123
85. Transverse measurement across head and great trochanter	132	148	158
86. Nutrient foramen to lesser trochanter	25	59	71
87. Diameter of head	49	59	59
88. Greatest transverse width of trochanter	53.5	61	65
89. Transverse width of external or lateral condyle	53	52	58
90. Transverse width of medial or internal condyle	40	50	50
TIBIA			
91. Length	380	430	430
92. Greatest width at proximal end	109.5	116	125
93. Width of middle of shaft	49	55	59
94. Greatest width at distal end	70	76	82.5
95. Antero-posterior measurement at proximal end measured from notch between articular surfaces from femur to top of crest	83	89	86

TABLE IX

PELVIS AND SACRUM

	Kouprey M.C.Z. 38108	Bantin A.M.N.H. 54470	Gaur A.M.N.H. 113147	Jersey Bull M.C.Z. 4
96. Diameter (transverse) of pelvic aperture	143	161	176	165
97. Diameter (sacro-pubic) of pelvic aperture	183	182	228?	170
98. Total transverse width of pelvic surface of the ilium	445	480	505	493
99. Pelvic surface of ilium from outer point of the wing of the sacrum to the tuber coxae	110	135	125	150
100. Pelvic surface of sacrum from outer point of the wing of the sacrum to edge of body of first segment	85	67	90	70
101. Dorsal side of ilium from gluteal line to highest point of the tuber sacrale measured on the perpendicular	130	130	155	145
102. Ventro-dorsal width of ilium shaft at narrowest point	48	56	69	50
103. Thickness of ilium shaft at above point	26	31	25	35
104. Length of sacrum on ventral side from anterior end of the body of the first sacral vertebra to posterior end of body of last sacral vertebra	207	239	260	265
105. Greatest total length of median crest formed by 5 fused sacral spines	198	210	237	225
106. Projection of dorsal medial crest of sacrum above lateral crest	80	75	100	55

COMPARISON WITH ALLIED GENERA

Elsewhere in this paper the differences and similarities between the hides and skeletons of the kouprey, gaur and bantin have been discussed. This section makes a general comparison of the skulls (without mandible) of the following living forms which are represented in each case by a single normal adult skull except that in the zebu and the Chillingham bulls more than one is used.

The skulls are of the kouprey, bantin, gaur, gayal, zebu (3), Holstein bull, Chillingham bulls (2), domestic yak, wild yak, wild water buffalo (Asia), wild Cape buffalo (Africa), and wild American bison. For the sake of uniformity the same skulls of the gaur and bantin are used that were compared with the kouprey in another section of this paper. The special subject of dentition is also discussed in another section and not dealt with here.

Table no. X (page 480) gives 29 comparative measurements of each of these 15 skulls and from these six indices have been derived to help to point out the relationship of the kouprey to this cross-section of various forms. The writer fully realizes that the measurements would be more valuable if based in every case on a large series but this would only have been possible in the case of the

American bison. Where several skulls were available an effort was made to select a good average male.

Among the skulls of varying size whose measurements are recorded the smallest is a zebu (A.M.N.H. 504). The next smallest are the two other zebus (M.C.Z.) followed by the domestic yak (A.M.N.H. 10730). These are of special interest for the morphological features of the skulls and the proportions between their various measurements.

The longest greatest length is that of a wild yak 104 mm. greater than the kouprey which is close to the bantin.

In basal length the wild yak is again the greatest with the kouprey closest to the gayal and Cape buffalo.

In palatal length the greatest is the water buffalo; the kouprey is closest to the Holstein and Chillingham bulls.

In length of nasals the gaur and wild yak have the longest. The kouprey is closest to a Chillingham bull and the American bison.

The greatest frontal length is that of the gaur and if we exclude the zebus that of the kouprey is the shortest. This is best shown by the frontal-length index based on the proportion of this measurement in each skull to its basal length x 100 (arranged in ascending order) kouprey 51, water buffalo 53, gayal 53, American bison 55, bantin 55, wild yak 56, zebu III 57, zebu I 58, Holstein bull 58, domestic yak 59, Chillingham bull II 60, Cape buffalo 62, zebu II 63, gaur 64, Chillingham bull I 70.

The absolute length of the palatal part of the premaxilla is greatest in the wild yak and next longest in the kouprey. The relationship is better shown by the premaxilla index based on the proportion of this measurement to the basal length x 100 (in descending order) domestic yak 32, kouprey 30, wild yak 27, Chillingham bull II 26, zebu III 26, Holstein bull 25, gayal 25, zebu I & II 24, gaur 24, American bison 23, Chillingham bull 22, bantin 21, Cape buffalo 21, water buffalo 21.

The length of the temporal fossa is greatest in the wild yak, that of the kouprey is closest to the water buffalo and one of the Chillingham bulls. The proportion of this measurement to the distance between the openings of the temporal fossae on the occiput x 100 has been used to determine the temporal fossa index arranged in ascending order: kouprey 45, gaur 56, wild yak 65, gayal 66, water buffalo 71, American bison 74, zebu I 76, Cape buffalo 78, bantin 79, domestic yak 80, zebu III 86, zebu II 103, Chillingham bull I 108, Holstein bull 119, Chillingham bull II 121.

The absolute length of the maxillary tooth row is greatest in the gaur and if we exclude the smallest zebu, it is shortest in the kouprey. The proportion of this to the basal length x 100 gives us a maxillary tooth row index arranged in ascending order: kouprey 26, bantin 27, wild yak 27, Chillingham bull II 27, gayal 28, water buffalo 28, Holstein bull 28, Chillingham bull I 29, zebu II 30, zebu III 30, Cape buffalo 30, domestic yak 30, American bison 31, zebu I 32, gaur 33.

No mandibular measurements are included in this comparison because the mandible of the kouprey is missing. It is hoped that this may be recovered and compared in a subsequent paper.

In greatest zygomatic width the gaur and the Holstein bull are the greatest if we exclude the zebus and the domestic yak. The kouprey is the narrowest.

In outside mastoid width the gaur is the greatest followed by Cape buffalo and a Chillingham bull. If we exclude the zebus and domestic yak, the gayal is the smallest followed by the kouprey.

In frontal width at orbits the American bison is greatest followed by the gaur. If we exclude two zebus the kouprey is least or narrowest.

In brain-case width the greatest is the American bison followed by a Chillingham bull. If we exclude the zebus the least is the kouprey.

In the width of maxilla measured outside the first upper molars, the water buffalo and the gaur are greatest. If we exclude three zebus the kouprey is the smallest.

In order better to compare the last five measurements a skull-width index has been determined by averaging the five width measurements and dividing this average in each case by the basal length. The resulting figure x 100 gives the following indices arranged in ascending order: kouprey 37.4, zebu I 37.6, zebu III 37.8, zebu II 38, wild yak 38, bantin 39, gayal 40, water buffalo 40, Holstein bull 42, domestic yak 43, Cape buffalo 44, gaur 45, American bison 46, Chillingham bull II 46, Chillingham bull I 49. It appears from this that the proportionately narrow skull is probably characteristic of the zebu as well as of the kouprey and is also shared by the wild yak.

Greatest combined-nasals width is greatest in the American bison followed by the wild yak. The least is the smallest zebu and the bantin. The kouprey is intermediate, closest to the water buffalo and a Chillingham bull.

Width of foramen magnum is greatest in the Cape buffalo. If we exclude the zebus and the domestic yak the next smallest is the kouprey in which it is the same as in the wild yak and a Chillingham bull.

Combined occipital condylar width is greatest in the wild yak. If we exclude the zebus and the domesticated yak it is least in the kouprey and water buffalo.

The width between the openings of the temporal fossae on the occiput is greatest in the Holstein and Chillingham bulls and even if we include all three zebus the smallest distance is found in the kouprey. This measurement is used in determining the temporal-fossa index already referred to.

The length of intercornual ridge (between bases of horn-cores) is greatest in the American bison followed by the gaur and a Chillingham bull. The shortest is probably that of the Cape buffalo where the horn bases almost meet. If we exclude this aberrant form and the smallest of the zebus' the shortest intercornual ridge is that of the kouprey and the next shortest that of another zebu. Strictly speaking there is no intercornual ridge in some of these forms as in the zebu and the kouprey. The measurement is taken between horn-core bases at the occipital part of the frontal where the ridge would be if it existed. An intercornual index has been determined by taking the proportion of this measurement to the basal length x 100 arranged in ascending order: Cape buffalo 9, zebu III 22, kouprey 26, Holstein bull 35, zebu I 38, wild yak 41, Chillingham bull II 43, gayal 48, zebu III 50, water buffalo 50, bantin 62, gaur 65, American bison 67, domestic yak 67, Chillingham bull 72.

In greatest width at the anterior end of the premaxilla the wild yak is widest with the water buffalo next. In this width the zebus are the least, followed by the bantin. The kouprey is intermediate similar to the gaur.

In the height from the palate to the upper tip of the nasals the gaur and American bison are greatest and if we exclude the zebus the kouprey and the water buffalo are least.

In the height from the upper rim of the foramen magnum to the occipital crest the greatest is that of the wild yak. If we exclude the smallest zebu the least is that of the gayal followed by the water buffalo. The kouprey is intermediate, closest to the bantin and American bison.

In the height from the upper rim of the foramen magnum to the frontal eminence the Cape buffalo is greatest followed by the gaur. The zebus are least, followed by the domestic yak. The kouprey is intermediate, closest to the wild yak.

In length of horn and horn-core the Cape buffalo is greatest followed by the kouprey and water buffalo. One would expect other specimens of the wild water buffalo to be greater than the kouprey. The kouprey horns are longer than those of the wild yak and longer than those of other true wild cattle in the forests of southeastern Asia. The smallest horns are the zebus' and Holstein bull's.

In the circumference at the base of the horn and horn-core the greatest is that of the Cape buffalo and next the water buffalo. The smallest are those of the zebus and Holstein bull. The kouprey is the same as the wild yak but less than the gaur.

The next three measurements have a direct bearing on what have been determined as progressive characters in the Bovinae.

The posterior angle of divergence of the central axis of the horn-cores from the frontal: These I have not in most cases measured to closer than within 5°. The greatest angle was 180° in the gayal and water buffalo. The smallest angle would have been in the anoa had we included this form. The smallest angle was 100° in the small zebu, next came the kouprey with 105°, and the next smallest the domestic yak with 135°, then the Cape buffalo at 140° followed by a zebu, wild yak, and the Holstein bull at 145°.

In the distance from the posterior orbital rim to the base of the horn-core the greatest distance is in the Holstein bull followed by the Chillingham bull. The primitive condition with the shortest distance is that of the water buffalo followed by the Cape buffalo. Following these comes the smallest zebu and then the gayal, kouprey and another zebu. The yak, gaur, and bantin are more progressive in this respect.

The projection of the premaxilla beyond the nasal is greatest in the Holstein and Chillingham bulls. It is least in the kouprey with the smallest zebu next.

In addition to comments on the measurements certain points have been noted in comparing the morphology of these skulls.

1. There was generally a small vomer not fused to the palate. The exceptions were the zebus and African buffalo that had a medium-sized vomer; the water buffalo had a large vomer fused to the palate.

2. Pilgrim has called attention to the variations in thickness of the palate posteriorly. With the exceptions of the zebus and the yaks which might be classed as medium the other palates were thick.

3. The projection of the central parts of the palate behind the maxillary tooth row is in part dependent on the age of the animal but nevertheless this condition is clearly more marked in some forms than others. In the 3 zebu skulls and the Holstein bull the palate did not reach as far back as a transverse line drawn behind the third molars. In the Chillingham bulls there was a projection of 5 mm. in the center of the palate, 11 and 12 mm. in the yaks, 30 mm. in the water buffalo and virtually no projection in the Cape buffalo and American bison. In the kouprey we find it to be 32 mm., similar to the water buffalo; in the gaur 10 mm. and in the bantin 10 mm.

4. By plotting a facial triangle with the line from the basion to the ethmoid foramen as a base it is possible to measure the angle of bending down of the facial part of the skull from the cranial axis. In this the kouprey is intermediate between the wild yak and the bantin. In ascending order the angles are gaur 119°, wild yak 124°, kouprey 132°, bantin 136° and zebu 140°.

5. Horn growth in cattle seems to be affected in a measure by the chemistry of the soil. This is discussed elsewhere in this paper (p. 476). In spite of factors which affect horn growth there appear to be definite patterns which characterize in some cases races, species, or genera of Bovinae. In the skulls we are comparing there are longhorn patterns, shorthorn patterns, and possibly mixed or hybrid patterns. The horn curve of the African Cape buffalo with heavy wide bases and fast tapering, up-curled tips is a definite pattern. The same may be said of the Asiatic water buffalo whether wild or domesticated with its widely bowed backwardly curving heavy horns. Among the wild cattle the longest horns are those of the kouprey and these have a characteristic curve described in another part of this paper. The curve pattern is quite similar to that of the wild yak and it may well be that the yak has retained this pattern from a kouprey-like ancestor. The patterns of the gaur and bantin are closer than many others but there appear to be essential differences. The gaur horns are much heavier and curve up in an open half circle whereas the more slender and rounded bantin horns are curved in and downward at the tips making a small but more nearly closed circle when seen from front or back. The short-horned wild form is the bison with its up-curved horns. Fossil bison must have had considerably longer horns judging by the horn-cores. Among the domestic cattle some of the Chillingham bulls have fine round horns going slightly forward, upward and back at the tips. This is similar to the curve pattern of the wild Urus as shown in the Augsburg painting (Lydekker, 1912, Pl. 3). These cattle are supposedly descendants of Bos primigenius. The domestic yak has a sort of hybrid curve that may result from mixed blood of zebu and yak. Our specimen of the gayal has horns going almost directly out at right angles to the skull with up-turned tips. This suggests a combination of zebu and gayal blood. The zebu skulls and the Holstein one we have, all had short simple curved round horns judging by the horn-cores.

In Dr. J. Ulrich Duerst's comprehensive monograph on horns (1926) he mentions his early theory that giant hornedness must depend on the moisture of the air and the climate of lake areas. He quotes as examples the longhorned Watussi cattle of Victoria Nyanza. He mentions that Neumann opposed him with the theory that it was lack of lime in the soil that produced large hornedness.

He then describes the interesting experiment of Dr. Otto Kellerhals with domestic cattle and concludes that there is a definite relationship of growth of horns to the chemistry of the soil. An important factor is the hydrogen iron concentration, an increase in size of horn being associated with the acidity of the soil and the lack of phosphates. This may very well be a matter of importance in horn growth and might help to explain the softer, more easily frayed horn of the kouprey were it not for the fact that the bantin lives in essentially the same habitat, probably has similar feeding habits, and shows no sign of deficiency in the horn.

I was greatly interested in Duerst's illustration (1926, p. 112, Abb. 43) showing excessive horn growth in a Cambodian ox. The translated title reads "Boeuf des Stiengs from Cambodia, a bantin cross with domestic cattle from the marsh country of Asia. The giant hornedness depends as well on too little hydrogen concentration in the soil. In this case the effect of castration is added." This skull is reported to be in the Paris Museum and there appears to be good reason to doubt the information that accompanies it, for the illustration shows a skull that in many ways resembles a kouprey with the characteristic yak-like pattern to the curve of the horns. The horns are not frayed at the tips, they are more rounded at the base than in the kouprey, the orbital rims are flat as in Bos indicus, and the anterior nasal spines are different. Nevertheless the horn curve, narrowness of skull, and particularly the lack of intercornual ridge and angle of the horn-cores indicates that here we probably have a hybrid between some form of domestic cattle and the wild kouprey, not the bantin. It will be interesting to obtain further data on this specimen. Such a cross as I suggest would be a far more logical explanation of the magnificent horns than just laying it to the chemistry of the Cambodian soil (which is not marshy in the northern part) as suggested by Duerst.

5. Pilgrim has based his classification of fossil bovids largely on the shapes of the cross-sections of the horn-cores and especially if keels were in evidence. The variability in length of horns is most evident in cattle domestication. One has only to look at the Ankole cattle of Uganda or the Texas longhorn. The basic horn pattern just discussed seems to be more stable, although the Egyptians early found ways to distort it artificially as well as by selective breeding. The shape of the cross-section of the horn-core seems to remain virtually unchanged and because of this may give us a reliable character to use in connection with their taxonomy. If a fossil horn-core with a triangular cross-section is found there is a good chance that it belonged to a buffalo, a rounded horn-core of a certain size usually characterizes Taurina.

The cross-sections in our skulls show: a thin wedge with a long point posteriorly as the Cape buffalo, a flattened triangle with short face anterior, the water buffalo, rounded with no keels bison, bantin, zebus, Chillingham bulls, Holstein bull; an almost round oval, gayal and wild yak; a comma-shape with a very short ridge or keel in the domestic yak, an oval narrow anteriorly in the gaur, a flattened oval with prominent posterior keel extending out for 9 mm. on the horn-core in the kouprey.

6. The intercornual ridge is in many cases nothing more than the occipital edge of the frontal, flat on top sloping towards the horn-cores on each side with often a raised frontal eminence in the midline. Such is the situation with the bison, yaks, Chillingham bulls, zebus, and gayal. In the Cape buffalo the high horn-cores encroach on top of the frontals until they are only separated by a narrow valley. In the water buffalo the top of the hinder part of the frontal is dome-shaped with a large hump instead of a ridge between the horn-cores. In the bantin to some extent and more so in the gaur we find a heavy intercornual ridge extending across the top of the skull, and in the gaur overshadowing the entire occiput. In the kouprey there is no intercornual ridge or frontal eminence. The top of the frontal and parietal slopes back toward the occiput and extends on the same level to the horn-cores. One is most reminded of the condition in the water buffalo if the central dome could be removed. Except for greater shortening of the parietal and no temporal ridge the kouprey reminds us of this part of the skull of Leptobos.

7. The general shape or proportion of the entire occiput in terms of greatest width to greatest height is for the African and water buffaloes and the gaur and gayal rectangular transversely; for the bison, yaks, Chillingham bulls, Holstein bull, and bantin slightly rectangular transversely; for the kouprey and zebus just about square. This fits in with the general narrowness of the zebu and kouprey skulls.

8. The auditory bullae are large and much inflated in the bison and two buffaloes; in the gaur, bantin, gayal and one Chillingham bull they are medium size and inflated; in the kouprey and the yaks they are medium size and compressed, in the Holstein bull they are medium size and greatly compressed, in one Chillingham bull they are medium size and flattened almost to paper thinness, in the three zebus they are small, medium, and large and all inflated. In general, flattened bullae are spoken of as a progressive character but there is such variation it can hardly be considered reliable. One has for example only to contrast the condition in the two Chillingham bulls.

9. The paroccipitals are very wide and heavily ridged in the water buffalo, of medium thickness with a heavy ridge in the Cape buffalo, thick in the wild yak but broken, heavily ridged of medium thickness in the Holstein and Chillingham bulls; they are small with a slight ridge in the bison, zebus, and domestic yak; small without ridge in gaur and bantin, and thick with a heavy ridge in the kouprey.

10. Elsewhere in this paper we have shown that the contact of the premaxilla with the nasals is far more common in certain groups than had been previously thought. Nevertheless a long contact probably represents the most primitive condition and this we would find in the anoa, next comes the water buffalo with 30 mm. contact, the gayal with 14 mm., a Chillingham bull with 12 mm., the kouprey with 10 mm., another Chillingham bull with 9 mm., a zebu with 4 mm. Among those with an interval between the nasal and the premaxilla we find the space 10 mm. in a zebu and the Cape buffalo, 13 mm. in Holstein bull, 20 mm. in a zebu, 23 mm. in a domestic yak, 45 mm. in a bison, 50 mm. in a wild yak, 9 mm. in gaur, 10 mm. in bantin.

11. There seem to be three patterns of spines on the anterior part of the nasals. Type I, probably the most primitive, has two central spines close together without accessories, Type II is like the first with small accessory spines on the outer side, Type III has four spines with the two outside ones longer than the two central ones.

Type I includes kouprey, wild yak, bison, and Cape buffalo. Type II gayal, gaur, Holstein bull, Chillingham bull. Type III water buffalo, 3 zebus, and bantin.

12. Under general notes we should mention that in the bison, nasals are very wide and the parietal comes up for a short way on to the frontal eminence. In the wild yak there is a slight fused ridge in the midline of the frontal as in the kouprey; in the Holstein and Chillingham bulls the prominent intercornual ridge overhangs the occipital plane. The three zebus are of unknown breed. They have narrow skulls and flat frontals in one case sloping back towards the occiput as in the domestic yak. The domestic yak has a small keel curving for 100 mm. under and partly around its horn core. The skull is very light, probably on account of poor diet in the menagerie where it lived. Unfortunately little faith can be placed in captive specimens. There are indications of zebu blood in this animal. The gayal shows a curious widening of the lower nasals such as I have only seen in Lydekker's (1880, Pl. xiv) figure of Bos platyrhinus.

The important points about the kouprey have been dealt with elsewhere.

To sum up, this comparison shows the kouprey to resemble many different forms in different measurements with a strong tendency to be an extreme and not an intermediate form. Most noticeable are resemblances to zebu, yak (wild and domestic), water buffalo and bantin. In the six different indices that were calculated the kouprey was greatest in the premaxilla index except for the domestic yak. It was close to the wild yak.

In the frontal-length index the kouprey was shortest except for the zebus, and the water buffalo was next. Shortness is a primitive condition.

In the temporal-fossa index the kouprey was last and next came the gaur, thereby reflecting the more primitive condition.

In the maxillary tooth row index the kouprey was least and the bantin and wild yak next. Again this is a primitive condition. In the index compiled from five average essential skull widths the kouprey was least, followed closely by the zebus and wild yak. In the intercornual index except for the aberrant Cape buffalo and the smallest zebu, the kouprey is the least or shortest, and thereby the most primitive.

In four measurements that have a direct bearing on progressive characters the posterior angle of the horn-cores shows the smallest zebu and the kouprey to be most primitive with 100° and 105° respectively.

The distance from the orbital rim to the base of the horn-core is shortest and most primitive in the buffaloes followed by zebu, gayal, and kouprey.

The shortness of the lower muzzle or the projection of the premaxilla beyond the tip of the nasal is shortest in the kouprey followed by the smallest zebu.

The kouprey and water buffalo share an extreme lengthening of the center part of the palate behind a transverse line across the backs of M^3.

With regard to horns Duerst believes their excessive growth is due to the chemistry of the soil and he figures a skull of a long-horned bantin hybrid that probably is a kouprey hybrid.

The significant things about horns appear to be least of all length. The pattern is of importance, but most significant is their cross-section. This is unique in the kouprey with its posterior keel extending out 355 mm. on the horn core and 9 mm. deep 100 mm. out from the base.

The kouprey reflects the more primitive condition in simple central nasal spines, and the premaxilla contacting the nasals for 10 mm.

In most of the comparisons the kouprey skull bears its greatest resemblance to the zebu but it is the most primitive on the average of all the wild skulls we are comparing, except for certain features of the buffaloes. It has definitely an Asiatic Taurine form but does not resemble closely any well known bovid in many of its characters.

TABLE X

COMPARATIVE SKULL MEASUREMENTS OF LIVING BOVINAE

	Kouprey M.C.Z. 38108	Bantin M.C.Z. 36669	Gaur M.C.Z. 36670	Gayal M.C.Z. 6183	Bos indicus—Zebu I M.C.Z. 174	Bos indicus—Zebu II M.C.Z. 8387	Bos Indicus—Zebu III A.M.N.H. 504	Bos Taurus—Holstein bull M.C.Z. 3	Bos taurus—Chillingham bull I M.C.Z. 8389	Bos taurus—Chillingham Bull II M.C.Z. 8388	Domestic Yak A.M.N.H. 10730	Wild Yak Acad. Nat. Sci. Phil. 17311	Water Buffalo M.C.Z. 38768	African Cape Buffalo M.C.Z. 23033	American Bison M.C.Z. 11611
Locality	Fr. Ind. Ch.	Fr. Ind. Ch.	Fr. Ind. Ch.	India	India	India	India	U.S.A.	Eng.	Eng.	Circus U.S.	Tibet	Fr. Ind. Ch.	Brit. E. Afr.	U.S.A.
Sex	M.	M.	M.	—	—	—	—	M.	M.	M.	M.	M.	—	M.	M.
Age	O.Ad.	Y.Ad.	Y.Ad.	Ad.	Y.Ad.	Y.Ad.	Y.Ad.	Ad.	Y.Ad.	Ad.	Y.Ad.	Ad.	Ad.	Y.Ad.	Y.Ad.
Greatest length	470	465	520	455	409	455	370	545	480	539	447	574	510	490	520
Basal length	446	456	470	440	382	415	333	496	425	493	400	507	490	452	480
Palatal length	302	291	312	280	227	243	202	298	255	300	269	322	328	285	494
Length of nasals	202	177	230	154	—	176	132	220	170	192	166	230	182	190	210
Frontal length	230	252	300	235	220	262	199	290	300	298	235	295	260	280	235
Premaxilla length—palatal	132	100	113	108	92	102	86	125	97	130	127	139	103	95	112
Temporal fossa length	165	155	177	160	127	130	104	154	166	160	140	185	165	160	183
Maxillary tooth row length	118	134	156	127	126	126	103	143	124	137	121	148	139	138	150
Zygomatic width (greatest) A.	194	209	234	215	174	184	143	234	231	228	185	215	225	216	219
Mastoid width B.	227	238	287	219	179	200	150	274	248	277	184	237	248	278	234
Greatest combined nasals width	66	57	74	77	61?	71	56	81	60	75	76	83	63	57	100
Frontal width at orbit C.	175?	195	275	190	155	185	153	256	250	265	235	245	220	250	280
Brain case width D.	111	122	122	126	102	107	88	136	145	224	128	126	140	112	229
Foramen magnum width	35	40	40	40	27	31	27	42	42	35	32	35	38	45	37
Combined occipital condylar width	106	108	121	102	82	86	72	126	127	119	95	129	106	109	114
Width between temporal fossae on occiput	75	123	100	106	96	134	89	184	179	204	112	121	118	126	137
Length of intercornual ridge	118	285	305	210	145	210	74	172	305	213	270	220	245	42	320
Outside width at first upper molars E.	129	132	148	140	110	112	97	141	138	141	127	141	150	138	146
Greatest width at end of premaxillae	100	83	100	74	62	76	66	105	87	103	86	109	106	105	96
Height from palate to upper tip of nasals	133	147	173	132	118	128	93	147	147	155	134	154	133	152	173
Upper rim of foramen magnum to occipital crest	84	82	88	66	70	75	58	90	92	85	75	100	71	87	86
Upper rim of foramen magnum to frontal eminence	160	176	184	129	98	101	78	145	130	134	106	149	182	190	129
Length of left horn	810	552	630	490	—	—	—	305	795	490	—	775	810	820	—
Length of left horn-core	553	412	467	255	151	—	129	220	—	—	450	—	—	—	235
Circumference at base of left horn-core	268	252	290	218	138	204	143	185	230	210	220	295?	325	—	200
Circumference at base of left horn	359	310	392	265	—	—	—	200	277	218	—	360	380	550	—
Angle of horn-core's divergence from frontal	105°	155°	170°	180°	160°	145°	100°	145°	175°	155°	135°	145°	180°	140°	155'
Orbit to base of horn-core	75	100	88	74	75	96	68	127	112	125	85	90	34	41	85
Maxillary projection beyond nasal	70	96	85	95	—	82	72	82	80	95?	105	115	89	69	103

COMPARISON WITH ALLIED FOSSIL GENERA

MORPHOLOGICAL DISCUSSION

The primitive characters of the kouprey skull and the fact that it is clearly not a close relative of any of the living bovids although it has points of resemblance with the living representatives of the genera Bibos, Bos, and Poephagus, lead us to a comparison with the fossil Bovidae. We are most fortunate in having available the recently published (1939) comprehensive monograph on the Fossil Bovidae of India by the distinguished paleontologist, Mr. Guy E. Pilgrim (Pal. Ind. N. S.). A great many of the fossil bovids described in the paper have been found in various horizons of the Siwalik Hills of Northern India, which are only 3200 km. in an airline from the present habitat of the kouprey.

Pilgrim does not follow the extreme course taken by Blanford (1891, p. 483) and Lydekker (1913, p. 11) of uniting all the Bovinae under the single genus Bos according smaller subdivisions merely subgeneric rank. He inclines to follow Rütimeyer (1878, p. 189) recognizing separate genera, but he groups Rütimeyer's Taurina, Bibovina and Bisontina under the Taurina provisionally, with Platybos in addition. For the purposes of this paper I shall adopt Pilgrim's classification in which he considers the following genera of Bovinae as valid, arranged in sections:

Bubalina: Parabos, Proamphibos, Hemibos, Anoa, Bubalus?, Bucapra
Syncerina: Syncerus
Leptobovina: Proleptobos, Leptobos
Taurina: Bison, Bibos, Bos, Poephagus, and Platybos.

Pilgrim emphasizes the important part that the shape of horn-cores of fossil cattle plays in helping to determine their phylogeny. He also separates Syncerus, the African buffalo, from the lineage of Bubalus or Proamphibos because the vomer is free from the palate which is obviously the original or more primitive condition.

In this section we shall first summarise those characters which Pilgrim considers primitive from his wide knowledge of fossil Bovidae and with special reference to what he calls the "evolution of individual characters in the Bovinae." From these we shall see in what respects the kouprey which we are studying is progressive and which of its features are primitive. We shall then indicate points of resemblance between a number of different described fossil bovids and the kouprey, and in conclusion determine the probable place of this new wild ox in part of Pilgrim's chart showing his suggested phylogeny of the Boselaphinae and Bovinae (see chart, p. 494).

Pilgrim emphasizes the fact that "our estimate of affinity in the Bovidae should be based on the sum total of characters not on one or two merely." He also points out "how indispensable for an accurate estimate of the phylogeny of a group is a knowledge of the continuous succession of genera and species through many geologic periods. Such a chain of evidence seems to me to be realised more completely in the case of the Boselaphinae and the Bovinae than in that of most other Bovid groups, so that some guarantee seems to exist that the conclusions at which I have arrived as to the taxonomy of these groups and as to the development of their special characters are not so far from the truth."

CROSS-SECTION OF HORN-CORE

"In the Middle Miocene the only antelopine horn-cores of which we have any knowledge possess a more or less elongately oval cross-section at first without keels, but later developing two keels antero-internal and postero-external which may be known as the two primary keels. The third keel which characterizes the Boselaphinae and the Bovinae originates merely as an asymmetrical growth either on the inner or the outer side of the primary keel axis. . . . It is not until we reach the stage of typical Bovinae, such as the Hemibos, that the third angle really becomes a keel. 'Pari passu' the two primary keels tend to lose their importance, especially the postero-external one and to approach one another so that the transverse much exceeds the antero-posterior diameter. Both these features become more marked in the genus Bubalus, although the antero-internal keel still remains moderately prominent. . . .

"I can scarcely avoid the conclusion that evolution of a kind similar to that which we know has occurred in the lineage of Bubalus is responsible for the oval, transversely expanded horn-core of Bison, Bibos and Bos." He then quotes proof of this by the two fossil Indian species of ox known as Bos acutifrons and Bos namadicus, by Pleistocene species of Bison and by Leptobos. "Both the species of Bos just referred to possess, at any rate occasionally, a well marked internal keel."

". . .Leptobos of which the affinity to Bos is suggested by its other skull characters, exhibits traces of the two primary keels, although the cross-section is almost circular with little lateral expansion. Unfortunately a horned skull of Proleptobos is unknown, but I should expect its horn-cores to have the two primary keels more pronounced than in Leptobos and perhaps an indication of the inner keel, on the assumption that the latter has been lost in Leptobos."

Pilgrim places great importance on the horn-core section and the location of the keels in determining the phylogeny. He has a plate of half-size cross-sections of nineteen different forms which tend to show that the more primitive forms had more emphasis on the primary keels while the more recent forms are more rounded, or as in Bubalus have a more prominently developed internal keel.

The cross-section of the horn-core of the kouprey is different from all of those figured by Pilgrim. If the kouprey's marked external keel is one of the primary keels then it most nearly approximates the flattened oval cross-section of Sivaceros which is considerably smaller in size but indicates the very primitive condition similar to those of the Middle Miocene with two primary keels. Dr. E. C. Colbert inclines to agree with me that the posterior keel on the kouprey must be a primary one.

If, on the other hand, the horn-core of the kouprey represents an oval, transversely expanded structure which occasionally possesses a well-marked internal keel it bears no close resemblance to Bison, Bibos or Bos, but it resembles that of Syncerus caffer in which Pilgrim considers the sharpened back keel as the interior rather than one of the primary keels. The laterally compressed oval cross-section in the kouprey, the relation of the flat surfaces to the axis of the skull, and the very marked postero-external keel with traces of the antero-internal one on the opposite side, together with the fact that the skull shows many more primitive than progressive features, lead us to conclude that this horn-core would be according to Pilgrim's analysis the most primitive type, hitherto not described in the Bovidae but predicted by its presence in Sivacerus and Miocene antelopes.

The size of the horn-core in the kouprey might appear to be too large for such a primitive bovid. If we compare it with such more recent forms as the larger skulls of Bos acutifrons, Bos namadicus, and Bos primigenius we realise that it is not a gigantic form, nor is it of unusual size when compared with recent Bovinae.

TORSION OF HORN-CORE

"We may equally assume that the Bovinae which have horns twisted in an anticlockwise direction were derived from forms with untwisted horns. . . . In Proamphibos in fact we actually are acquainted with a type of untwisted horn-core which seems to have directly preceded those of Hemibos and Bubalus

. . . For the same reason neither Leptobos, nor even perhaps Proleptobos is likely to have been the direct ancestor of Bos, since the horns, at any rate of the former, are slightly twisted anticlockwise."

The horns of the kouprey are slightly twisted anticlockwise but not sufficiently to prevent their being ancestral to those of Bos, or Poephagus.

SITUATION OF HORN-CORES

"The shifting back of the horn-core from the primitive position immediately above the orbit is a progressive character found in dissimilar groups, but finds its supreme expression in the Bovinae. It is correlated with the lengthening of the frontal at the expense of the parietal."

"As early as the Pontian however a backward shifting of the horns has begun in the primitive buffalo, Proamphibos, and becomes a distinct feature in the Upper Pliocene (Villafranchian) Hemibos. On the lineage of Bos, Bos acutifrons of the Pinjor stage already has the horns even farther from the orbits than in the living buffaloes, and the distance is increased as time goes on."

In the kouprey the bases of the horn-cores are 75 mm. behind the orbits as compared with 90 mm. in the gaur and 110 mm. in the bantin. The kouprey is more primitive in this respect but not as markedly so as in some other points here discussed.

DIVERGENCE OF HORN-CORES

"The lateral projection, or as we may term it the downward bending of the base of the horn-cores, is one of the most marked characters of the subfamily of the Bovinae. . . . Even in the most primitive Bovines known as Proamphibos and Parabos the horn-cores do not diverge at an angle of more than 45°, but in the later form Hemibos, this angle has increased to 90°, while in the still more advanced Bubalus and Bos it generally approaches 180° and often exceeds this. For the Bovinae at any rate we may therefore regard this as a regularly progressive character."

The angle of the horn-cores in the Anoa which is in many ways a primitive type may be compared with the more familiar Bubalus as representing a progressive type.

The angle of divergence of the horn-cores in the kouprey (M.C.Z. 38108) is close to 105° which is not far from the angle of 90° in Hemibos and considerably less than in any living wild male Asiatic Bovinae with the exception of the Anoa.

FRONTAL

Pilgrim points out that in the Bovinae the breadth of the frontal has become accentuated with the increase in the size of horns. This can be traced from small-horned Proamphibos through Hemibos into Bubalus and from small-horned Proleptobos into Bos and Bison.

In the kouprey we have considerably larger horns than in the bantin and larger horns than in the gaur. We should therefore expect to find a wider frontal. This is not the case for the frontal width measured at the supraorbital process is, as shown elsewhere in this paper, proportionately similar in the kouprey and bantin and about 18% less than in the gaur. The frontal width at the orbits is not a satisfactory measurement due to individual variation in the supra-orbital process. The frontal width measured at the narrowest point of the postorbital constriction is in the kouprey about 7% less than the bantin and 27% less than the gaur. This shows that the kouprey in spite of its increase in horn size shows the more primitive condition of a narrower frontal.

In the Bovinae the frontal not only expands transversely but also longitudinally. "In Bubalus the frontal is often three times and in Bos four to six times as long as the parietal. Both in Tetraceros and Boselaphus as well as in the Bovinae the horn-cores move backward gradually, concurrently with the backward extension of the frontal, so that in Bos they come to lie over the plane of the occipital with which the much reduced parietals are fused." In the kouprey the frontal has a length of 230 mm. as compared with 260 for the bantin and 290 for the gaur with which we have compared it. This measurement naturally varies with the size of the animal but proportionately the kouprey represents the more primitive condition in a comparison with Bibos. The horn-cores of the kouprey are not placed as far back over the plane of the occipital as are those of the gaur and bantin.

SUPRAORBITAL FORAMINA

"In the Bovinae the supraorbital foramina have as a rule remained closer to what we assume was their original size. . . . Leptobos has broad and moderately deep supraorbital pits. The same condition has generally been retained in Syncerus. On the contrary the pits have lengthened and narrowed considerably in all the Pleistocene and recent forms belonging to the Taurina and Bubalina." In the kouprey the pits are of the more lengthened and narrow type similar to those in Bibos.

Pilgrim points out that the original position of the supraorbital foramina was opposite the anterior edge of the orbits and the pits were short. As the horns moved backward the foramina followed them leaving the long narrow groove. This change of position is probably also the reason for the narrowing and long thinning of the pits already mentioned. The foramina have increased in number forming a chain of openings and the elongate pits have sometimes become roofed over by bone. This can be traced in Leptobos, Hemibos and Bubalus but it reaches its extreme in the Taurine section. The dorsal view (Plate VII) of the kouprey skull shows prominent pits and grooves of the Taurine type. The width between the supraorbital foramina is 2.7% less than the bantin (M.C.Z. 36669) and 32.7% less than the gaur (M.C.Z. 36670). The pits are proportionately closer to the orbits in the kouprey than in the bantin and gaur.

LACHRYMAL

"In general the lachrymal seems to have deepened, especially in its anterior part, in the Bovinae and this character is equally marked in the genus Tetracerus.

"This may be correlated with the ossification which has apparently led to the gradual elimination of the ethmoidal fissure. Winge regarded this fissure as an advanced character but it is difficult to reconcile this with its universal presence in primitive families of the Tragulidae, the Cervidae, most of the Boselaphinae and primitive Bovinae, as against its absence in advanced Bovinae. . . . On the whole the facts do not seem to me to support Winge's opinion."

In the kouprey the lachrymal has a width of 30 mm. as compared with 40 mm. in the bantin and gaur. In this respect it reflects the more primitive condition. Pilgrim has just pointed out that the ethmoid fissure or vacuity is present in primitive families of Bovinae and absent in advanced Bovinae. In this he takes issue with Winge and we agree with Pilgrim. The presence of a prominent ethmoid vacuity in the kouprey is I believe the first recorded case among the living wild Bovinae. The fact that it is present in certain fossil bovids (notably Leptobos and Proleptobos) and that so many of the characteristics of the kouprey are primitive makes it less surprising in this form than it would be in a more progressive bovid.

Pilgrim says that "Proleptobos shows a trace of a very shallow lachrymal fossa, but this has vanished in Leptobos. In none of the later nor living Bovinae is a lachyrmal fossa present." The kouprey has no trace of one.

PREMAXILLA

"The premaxillary region is not preserved in any of the Indian fossil Boselaphinae, but judging by the living Boselaphus and Tetracerus it seems certain that the premaxilla extended far back in the subfamily as a whole and that it had a very considerable union with the nasal. In Boselaphus it generally does not reach as far back as P^2, though somewhat variable. In Tetracerus its hinder end is far in front of P^2.

"A similar condition is plainly visible in the early Bovine Hemibos, where the premaxilla extends back quite to the level of P^2 and has a long union with the nasal, as is the case in the living genus Anoa. On the other hand, in both Bubalus and Bos the premaxilla ends in front of P^2 and has no union whatever with the nasal."

In this statement Pilgrim clearly indicates that the primitive condition is that in which the premaxilla has a considerable union with the nasal. Such is the case in the kouprey where the union is not proportionately as great as in the Anoa but there is a definite contact for a width of 10 mm. in front of P^2. We cannot agree with Pilgrim's statement that "in both Bubalus and Bos the premaxilla ends in front of P^2 and has no union whatever with the nasal." Cossar Ewart has described a number of skulls (ref. P. Z. S. 1911) where such a union takes place. "In the buffalo the total length of the premaxilla may be 173 mm. and its connection with the nasal is 35 mm. . . . the Indian Buffalo lives in the neighborhood of swamps and jungles, and probably also feeds on reeds and coarse grasses—food which necessitates long, firmly secured premaxillae. . . . In Bos primigenius the premaxillae in all the skulls I have examined reach the nasals. In some cases the connection with the nasals is only 5 mm. . . . In some of the skulls from the Newstead Fort the premaxillae are large and have nearly as extensive a connection with the nasals as in the Buffalo; in others they are short and separated from the nasals by a considerable interval . . . the premaxillae probably reached the nasals in at least 90 per cent of the skulls of the Celtic shorthorn (Bos longifrons) type and in about 70% of the skulls belonging to the long-horn and cross-bred animals."

Such a union is not considered characteristic of Bibos although Cossar Ewart and Pocock have both reported it in a gaur skull and we have a bantin skull in which it occurs. It is also evident in certain races of domestic cattle notably the Holstein bull, Chillingham cattle skulls and a specimen of Bos indicus. In the bison there is a wide interval between the top of the premaxilla and the nasal while in Poephagus the nasal process of the premaxilla is very slender but it generally does extend as far as the nasal.

PARIETAL

Pilgrim points out that in the lineage of Syncerus the parietal ends (anteriorly) in a narrow tongue which does not reach the orbit. "A similar narrow tongue represents the parietal in the front part of the temporal fossa in the case of the genera Poephagus, Bibos, Bison, and Bos. In the general area covered by the parietal Bibos and Poephagus are the most primitive of these genera and Bos is the most advanced. In the most progressive of all these genera, Bos, two additional developments have taken place. These are: the fusion of the supraoccipital with the parietal and the entire deflection of both into the plane of the occipital. Several intermediate stages are observable between this extreme condition and that which exists in the Boselaphinae and the antelopes generally. Thus in Bibos in the juvenile stage not only is the suture between the parietal and the supraoccipital quite open, but the parietal roofs over the braincase almost completely. Even in the juvenile stage of Bos, the parietal forms an obtuse angle with the occipital, although the suture between the parietal and the supraoccipital is almost invisible."

In the temporal fossae the kouprey's parietal terminates in a narrow point on a line with the front of the horn-core and the foramen ovale. This does not extend as close to the orbit as in Bibos, Poephagus, Bison, and Bos indicus, although there are other races of Bos that show a similar parietal structure (see Fig. 132, p. 134, Sisson).

TEMPORAL CRESTS OF PARIETAL

Pilgrim points out three main factors with which are bound up the origin and development of the temporal crests. These are applicable in the case of the primitive Bovinae such as Proamphibos, Parabos, Proleptobos and Leptobos. The need for them is much less in the later Hemibos where the face is more bent down on the occiput, the frontal has lengthened and there is a strong transverse ridge between the horn-cores. "In the still more advanced genera Bubalus, Bibos and Bos this alteration in the structure of the skull has proceeded to an extreme degree, and all trace of the temporal crests has vanished." In the kouprey they have also vanished but the plane of the parietal forms an obtuse angle (30°) with the occipital and is not in the same plane (Pl. XI, fig. 2). In the bantin the parietal part of the occiput forms part of the occipital plane, in the gaur the heavy intercornual ridge forces it to slope back. This is also the case to a less degree in Bison, Poephagus and various races of Bos. The occipital part of the parietal in the kouprey suggests more that of Syncerus and Bubalus.

OCCIPITAL BONES

Pilgrim says, "In the Bovinae the widening of this part of the skull is still more remarkable but it is manifest that in this subfamily it gradually comes to embrace the whole of the occipital."

"There is however a disproportionate increase in breadth between the upper and lower portions of the occipital. Thus in the group of the cattle including Bison, Poephagus, Bibos and Bos the extreme flattened semicircular shape of the occipital in modern species of Bos has taken the place of a triangular shape in the Upper Siwalik Bos acutifrons, while in Poephagus the occipital possesses an even more triangular shape in which I seem to see the survival of a form still more primitive than Bos acutifrons."

This last statement seems almost prophetic as far as the kouprey is concerned for it possesses a comparatively narrow and almost triangular-shaped occiput (Pl. XI) back of the horn-core. Pilgrim says another character of considerable systematic importance concerns the relation of the temporal fossae to the occipital. "Simultaneously with the backward movement of the frontals, the parietals and the supra-occipital are abbreviated and deflected into the plane of the occipital, so that the temporal fossae open entirely on to the occiput. The general widening of the latter region then tends to increase the spacing between the ends of the fossae. This can be observed both in the Bubalina and the Taurina as exemplified by Bubalus as compared with Bibos. On the other hand, in the group of the Leptobovina the hinder ends of the temporal fossae have not shared in the widening of the occiput but have become curved, so as to maintain or even increase their proximity to one another. The development of Leptobos from Proleptobos clearly shows how this may have occurred. This striking difference of the Leptobovina from the other two Bovine groups in the development of the temporal fossae militates against the possibility that Proleptobos was ancestral to the Taurina."

The kouprey clearly does not represent the Leptobovine occiput. It has however temporal fossae opening so prominently on to its occiput that there is a space of 75 mm. between the ends of the temporal fossae as compared with 100 in the gaur and 123 in the bantin. As Pilgrim has indicated, Bibos is more primitive than Bos in this respect. Most of the skulls of Bos primigenius have the temporal fossae completely excluded from the occiput on account of a bony plate, formed by the extreme widening of this region of the skull. The condition in Poephagus is somewhat intermediate between Bos and Bibos in this respect.

BASICRANIUM

"The widening of the skull at the mastoid is a special characteristic of the subfamilies of the Boselaphinae and Bovinae, which began to manifest itself from a comparatively early period. Correlated with this is a progressive increase in the size of the occipital condyles, and the foramen magnum, and in the breadth of the basioccipital, particularly in its hinder part." In the kouprey as compared elsewhere in this paper with the bantin and the gaur we find a mastoid width 2% less than in the bantin and 16% less than in the gaur, an occipital condylar width similar to that in the bantin and 7.5% less than in the gaur. In these two respects the bantin and kouprey were essentially similar. In the width of foramen magnum the kouprey was 10% less than the bantin and 7.7% less than the gaur.

Pilgrim says the primitive structural shape of the basi-occipital is short with a triangular outline. Again broadening and flattening of the whole surface takes place in the Taurina and is particularly marked in Bos and Bison. In this respect the kouprey basi-occipital is more primitive, like that in Bibos although narrower at the condylar end. In discussing the auditory bulla Pilgrim says, "The bulla is much smaller and narrower in Bubalus than in Hemibos and still more than in Proamphibos, but the climax of reduction is reached in Bos, in which the bulla though lofty is almost plate-like. On the other hand in the Pinjor genus Bucapra the bulla is quite as broad as in Proamphibos and is much more inflated. Somewhat the same primitive condition remains in the living genus Syncerus. With the widening of the skull at the mastoid the bulla assumes a more oblique position with reference to the sagittal axis. As might be expected this condition becomes more marked in the Bovinae."

The bulla in the kouprey is distinctly flattened and laterally compressed resembling that of Poephagus, the wild yak, or in some cases, Bos indicus rather than the more inflated bullae of the bantin and gaur. According to Pilgrim's statement this is a progressive rather than a primitive character. It is the first of those which we have discussed in which the kouprey has not shown the primitive condition. In 1931 Van der Klaauw wrote a comprehensive paper on "auditory bullae in fossil mammals." From this I gather that the degree of inflation of the auditory bullae in the Bovidae is subject to considerable individual variation. The American Museum has two domesticated yak skulls, one of which shows an inflated bulla and the other a flattened one.

Pilgrim points out that in primitive Bovinae the cross-section of the paroccipital process is rather stout and triangular. In later Bovinae it is quite as narrow laterally as in Boselaphus. "We may therefore conclude that probably the general trend of evolution was from a stoutly built, circular or equilateral triangular paroccipital process to one that was greatly compressed laterally."

The kouprey has a stoutly built paroccipital or paramastoid process which is heavily ridged and squared on the ends. In Bibos these processes are much slenderer. This adds another probably primitive character to the kouprey skull.

PALATE AND VOMER

"Correlated with the progressive increase in bulk of the skull and horns in the Bovidae is the formation of cancellous tissue within certain bones. It is particularly marked in the frontals and horn-cores of the Bovinae and has long been known and been taken as one of the characters of that subfamily. The development of similar sinuses in the hard palate of the Bovinae accompanied by a thickening of the bone, especially pronounced in the median sagittal line, has not, however, so far as I am aware, received any comment."

Pilgrim says that such a thickening of the palate does not occur to his knowledge in any of the Boselaphinae, but in one of the earliest and perhaps the most primitive of the Bovinae, Proleptobos birmanicus, it is already fully developed although less so than in its successor Leptobos and much less so than in the later oxen as well as in the buffaloes. In an excellent plate (No. VI) he shows vertical cross-sections of the palatal structure just behind the last molar in eleven different Bovids.

Of those shown the kouprey most closely resembles in shape that of Bos longifrons. In thickness it is similar to specimens of Bos, Bison, and Bibos but thicker than in Poephagus. It is unfortunate that Pilgrim's plate only included the Bison among the living Bovinae although the similarity of Bos, Bison and Bibos is great in this respect. The palate thickness in Poephagus suggests more that of Syncerus.

Pilgrim's palate cross-sections also indicate those forms in which the vomer is upwardly extended until it becomes fused with the palate. Such a vomer is evident in Bubalus but not in Syncerus. It is absent also in Proleptobos, Leptobos and in each one of the genera included in the section Taurina. "On the other hand in Proamphibos, the earliest known member of the Bubalina, the fusion of the vomer and palate is easily visible. This seems to afford conclusive evidence that Proamphibos is not ancestral to the Taurina or Syncerina as it undoubtedly is to the Bubalina, but that the three sections diverged, probably as early as the Sarmatian, from a common ancestor in which the thickening of the palate was probably less and in any case union with the vomer had not taken place."

The kouprey definitely belongs with the Taurina in this respect and might well qualify as a "contemporary ancestor for them" as its vomer is small and not united with the palate.

DENTITION

Pilgrim points out that primitive characters in the dentition of the Bovinae may be:

1. "The strong development of the folds and ribs on the external side of the upper molars and the convexity of the internal surface of each lobe of the lower molars." Excessive antero-posterior compression of the molars has produced median ribs of extraordinary strength both on the inner and outer surfaces of the external lobes. Pilgrim points out that this development of strong median ribs is the beginning of an advanced development in the Bovinae.

2. "The comparatively small degree of reduction of the premolars" is another primitive character of the Bovinae as is "the retention of the primitive structure of P^4." The Taurina show a more intense antero-posterior compression of P^4 and a greater diminution in the size of P^2.

There is a gradual lengthening of the molar series and of the individual molars. "The primitive quadrate shape of the upper molars can be observed not only in Butragus but also in the Upper Miocene and Pontian Boselaphinae and in the living Tetracerus."

"In the Bovinae a progressive advance in the same direction is more pronounced. The quadrate shape of the upper molars in Proleptobos and Proamphibos and equally in the living Anoa is characteristic of these genera. Some elongation has taken place in Leptobos and Hemibos and in a much less degree in Bubalus. In Bos and still more in Bison, the upper molar series has lengthened considerably."

Pilgrim lists measurements of dentition in the following table to which we have added for comparative purposes those of the kouprey, the bantin, and the gaur.

TABLE XI

MEASUREMENTS OF DENTITION IN BOSELAPHINAE AND BOVINAE

(Modified from Pilgrim)

	Boselaphus tragocamelus	Proamphibos lachrymans	Hemibos acuticornis	Proleptobos birmanicus	Leptobos falconeri	Bubalus bubalis	Bos taurus	Kouprey	Bantin	Gaur
Length of premolar series	48	60	59	53?	61	75	58	50	58	61.5
Length of molar series	69	84	82	77	82	97	85	70.5	76	94.5
Ratio x 100	69.5	71.4	72	69	74	77	68	70.9	76.5	65
M^3 Ant. post. diam.	26	30	32	28	32	34	28	25.5	26	36
M^3 Trans. diam.	—	27.5	27	29	28	—	—	23	19	26
M^3 Height of crown	37	45	—	—	—	78	45			
Ratio height: ant. post. diam.	142	150	—	—	—	220	161			

From this table it is evident that both the upper premolar and molar series are short in the kouprey approaching most closely that of Boselaphus tragocamelus and among the Bovinae that of Proleptobos birmanicus. The anteroposterior diameter of M^3 was closest to Boselaphus. The narrowness of the premolars in the kouprey suggests that of Leptobos elatus, and Leptobos falconeri, both primitive forms.

The ribs on the external side of the kouprey's upper molars are not as strongly developed as in other recent Bovinae. The premolars probably retain their primitive length. Unfortunately the mandible is missing but upper P^4 shows an antero-posterior compression. The kouprey's upper molars show the same primitive quadrate shape so characteristic of Proleptobos, Proamphibos, and the living Anoa. Later Bovinae show extreme hypsodonty and enlarged basal pillars on the inner lingual side of the upper and lower molars. The kouprey is also primitive in this respect without the enlarged basal pillars.

METAPODIALS

"Data as to the structure and proportions of the limb bones of the fossil Boselaphinae are so fragmentary, that no solid conclusions can be drawn as to their evolution along various lineages. In fact this section would have been altogether omitted were it not for the remarkable change which the metapodials have undergone in the Bovinae. It seems probable that the metapodials were ong in all the primitive Boselaphinae. . . . The shortening of the metapodials is no doubt to be correlated with the increased mass of the animal. At any rate it can hardly be doubted that this increase in bulk has been largely responsible for the stoutening of the limbs and foot bones and the shortening of the metapodials in the Bovinae. . . . Correlated with the stoutening of the feet, the lateral expansion of the extremities of the metapodials is very pronounced in the Bovinae. . . . The Pleistocene members of the group of the Taurina display the character very strongly."

The metapodials of the kouprey are more slender and less expanded at the extremities than those of the gaur and bantin with which they have been compared in another part of this paper. The whole leg shape of the kouprey is more slender than that of other living Bovinae of equal or greater size.

PHYLOGENY

Pilgrim lists his subfamily Bovinae with which we are here concerned as containing four groups of genera—Bubalina, Syncerina, Leptobovina, and Taurina.

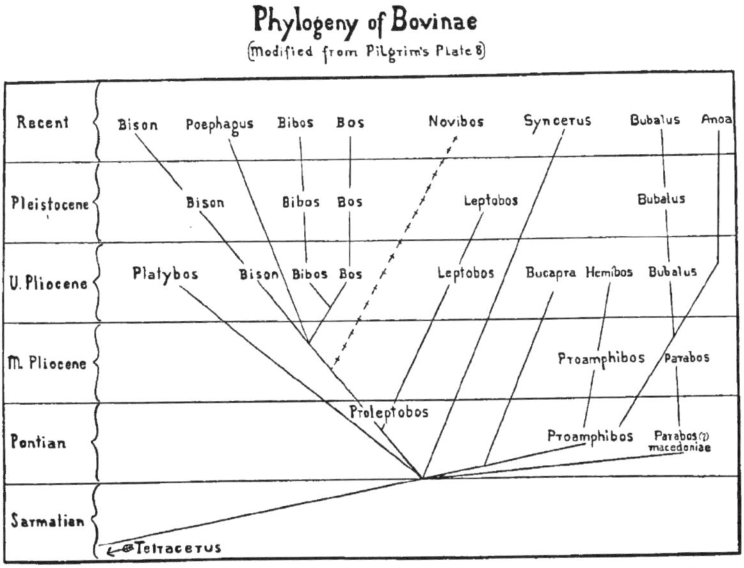

Fig. 9. The dotted line on this Phylogenetic Chart shows the probable relationship of Novibos to other living and fossil Bovinae, as shown in this memoir.

In discussing his phylogenetic chart, a part of which we will take the liberty of reproducing, he states, "The horn-cores I may say, form the basis for the classification as a whole, not only because they are generally for fossil forms the most easily accessible part of the anatomy, but also because their very plasticity renders them most suitable as an indicator of the divergences from the primitive stem which have taken place at different times." Other anatomical factors also contributed to his reasons for the charted phylogeny.

The comparisons that we have made would indicate that the kouprey, which is distinctly a different genus from Bison, Poephagus, Bibos, or Bos, is nevertheless closest to the generic groups of Taurina; although considerably more primitive than other living genera, it is not as primitive as Proleptobos and we therefore branch it off from the parent stem of the Taurina sometime in the Middle Pliocene. It bears a closer relationship to the Taurina than Syncerus does to the Bubalina. (see fig. 9.)

TABLE XII

MEASUREMENTS OF AMERICAN MUSEUM BOVINE SKULLS FROM THE UPPER SIWALIKS (after Pilgrim)

	Leptobos falconeri A.M.N.H. 19816	Hemibos acuticornis A.M.N.H. 19963	Bubalus platyceros A.M.N.H. 19872	Novibos Sauveli M.C.Z. 38108
Width of skull at mastoid	217	app. 190	269	227
Width of skull at orbits	227	app. 200	265	175
Width of skull at constriction above orbits	213	app. 185	240	170
Width of skull between temporal fossae	124	app. 132	143	111
Distance from occipital crest to highest point of frontals	127	app. 97	113	87
Distance from highest point of frontal to apex of nasals	app. 117	app. 118	242	157
Distance from apex of nasals to end of premaxilla (estimated)	app. 240		app. 250	259
Length of temporal fossa	134	app. 200	190	165
Interval between supraorbital foramina	92	app. 95	141	99
Length of horn-core measured from pedicle along convex curve (estimated)	590	400	880	553
Antero-posterior diameter of horn-core at base	77	55	98	104
Lateral diameter of horn-core at base	85	75	146	62
Interval between orbit and base of horn-core	100	64	50	75

TABLE XIII

COMPARISON BETWEEN LEPTOBOS FALCONERI AND NOVIBOS

	A.M.N.H. 19816		M.C.Z. 38108	
1. Height of occiput from base of condyle	133		142	
2. Distance between openings of temporal fossae on the occiput	83		75	
3. Distance from orbital rim to base of horn-core	85		75	
4. Contact of premaxilla and nasals	20		10	
5. Projection of center of palate behind tooth row	33		32	
6. Shape of pterygoids behind palate	splayed out		splayed out	
7. Width of palate inside molars	87		83	
8. Combined length of molars	85		70.5	
9. Combined length of premolars	54		50	
	A.P.	L.L.	A.P.	L.L.
10. Fourth upper premolar widths	17	29	16	21
11. M^1 widths	24	29	20	22
12. M^2 widths	25	25	23	23
13. M^3 widths	35	21	25.5	23

TAXONOMIC DISCUSSION

Proleptobos and Leptobos both have ethmoid vacuities as has the kouprey. The teeth in Proleptobos are less elongate, the upper molars being quadrate and the premolars broader than in Leptobos. In this they more nearly resemble the kouprey.

The length of upper molar series in Proleptobos birmanicus (Br. Mus. No. M. 10909) is 78 mm. while it is 70.5 in the kouprey. The upper premolar series is approximately 54 mm. compared with 50 mm. in the kouprey.

Pilgrim's Proleptobos also resembles the kouprey more than Leptobos does in the relatively more narrow skull than that of Leptobos; in the relatively broader face and relatively greater height of the skull; in the smaller supraorbital pits. The kouprey resembles Leptobos more in the closeness of the opening of the temporal fossae on the occiput and the absence of a lachrymal fossa. For comparison of the kouprey measurements with Leptobos falconeri, Hemibos acuticornis, and Bubalus platyceros see tables XII and XIII.

Duerst (1926) ranks Leptobos with the antelopes but Stehlin (1927, p. 217) states that Leptobos in the shape of the tooth rows and as far as the shape of the skull is concerned as well as the extremities of the skeleton comes closer to the bovids especially to the bantin much more than to any recent form of antelope. Pilgrim also classes Leptobos with the cattle rather than the antelope.

Pilgrim says, "It seems to me that Bibos gaurus and Bibos banting both in the embryonic structure of the parietal as explained by Stehlin (1893, p. 33) and Duerst (1905, pp. 235, 252) in the shape and position of the horns, and in the deeper and more open temporal fossae represent a less progressive type than either living or extinct forms of Bos taurus or the Pleistocene Bos primigenius. This view is borne out by our knowledge of the development of the same features in other Bovine lineages. Thus Proamphibos, Proleptobos and Bison sivalensis, considered as early stages of Bubalus, Leptobos and Bison bison respectively, afford structural analogies to Bibos in its relation to Bos. Naturally we may expect to find that ancestral forms of Bos will tend to approach Bibos more nearly, the farther back we go in the geological record, and although this is actually so in the case of the Siwalik species mentioned (Bos acutifrons, Bos planifrons, Bison sivalensis, and Bos namadicus), yet it seems to me impossible that species which have already progressed some way towards the development of characters which distinguish Bos from Bibos could be ancestral to Bibos (see p. 323). For this reason I consider that the more natural position of the Indian fossil species in question is with 'Bos sensu stricto.' The living species of Bibos appear to be survivals of a form which existed sometime previous to the Pinjor stage of the Upper Siwaliks, which would be the common ancestor of Bos and Bibos."

As pointed out in this paper there are many characters of the kouprey which combine those of Bibos and those of Bos which would qualify it as a close relative of the common ancestor referred to by Pilgrim.

In comparing Bos namadicus and Bos primigenius Lydekker in his later writings says that "typically it is very closely allied to the latter species, but in some instances the horn-cores are more or less flattened at the base, and thus approaching to the bibovine type."

Pilgrim points out that the great expansion of the nasals at their infra-orbital angle is a Bos characteristic whereas in Leptobos the nasals taper regularly from their distal end upward.

He compares certain characters in which Bibos has retained the primitive condition while Bos namadicus has attained very nearly to that which occurs in the most progressive type of Bos and from this he concludes that namadicus lies some distance from the point where the two lineages of Bos and Bibos diverged.

1. "Longitudinal extension of the frontals which greatly exceed in length the part of the face anterior to them. In Bibos the two dimensions are approximately equal." In the kouprey the two dimensions are approximately equal.

2. "Occipital crest reaches above lower border of the horn-cores so the proportion of the occiput above it is relatively small and the parietals and supra-occipital which are comprised in it are reduced to an extremely narrow strip. In Bibos the summit of the occipital crest lies far below the lower border of the horn-cores and above it the parietals and supra-occipital occupy a relatively wide band."

In the kouprey the summit of the occipital crest line reaches just above the lower border of the horn-cores.

3. "The temporal fossae open by narrow slits on to the occiput only just above the occipital crest their hinder ends being distant from one another by an interval which often is nearly double the width of the occipital condyles. In Bibos the temporal fossae open widely on to that portion of the occiput which extends far above the occipital crest, and includes both the supra-occipital and the parietal; also their posterior extremities approach one another more nearly."

In the kouprey the opening of the temporal fossae on the occiput is not as wide as in Bibos but it is of a similar type only there is not a large portion of the occiput extending above it. The posterior extremities of the temporal fossae approach one another considerably more than in Bibos.

We can further profit by Pilgrim's comparison of Bos namadicus and Bibos as applied to the kouprey.

Pilgrim lists certain "characters in which Bos namadicus agrees with Bos and differs from Bibos but in which Bibos is not obviously primitive and may indeed be more specialized:

4. "Excess in length of the antero-posterior over the transverse diameter of the frontals, just the reverse being the case in Bibos." Kouprey agrees with Bos in this.

5. "Horn-cores extending far in front of the face and with their upper border near the base convex. In Bibos the horn-cores do not extend in front of the face and their upper border is concave." Kouprey is intermediate in this with the upper horn-core border convex but the horn-cores not extending in front of the face.

6. "Nasals long and tapering, widest at the infra orbital angle. In Bibos they are short and widest at the distal end." In kouprey they agree with Bos.

On the other hand Bos namadicus approaches Bibos:

(1) "In the frequent retention of an ovate cross-section of the horn-core, thereby showing a vestigial keel, which has vanished in the most extreme form of Bos." The kouprey's horn core cross-section is also oval with a prominent keel.

(2) "The premaxillae are generally remote from the nasals, whereas in the most advanced forms of Bos, they are in contact with the nasals." They are also in contact in the kouprey a more primitive condition.

To sum up, Pilgrim points out that many of Lydekker's differences between Bos namadicus and Bos primigenius are not constant but he inclines to believe that the great extent to which the supracristal overhangs the infracristal portion of the occiput is peculiar to the Indian species. Also the ovate cross-section of the horn-core is rare in Bos primigenius. The twisting anticlockwise in primigenius is more even than in namadicus and the horn-cores extend to a much less degree in front of the face than in the Indian form. In Bos primigenius the premaxillae are closer to the nasals than in Bos namadicus. The occipital tends to be somewhat higher in Bos namadicus than in Bos primigenius and the temporal fossae tend to indent the occiput somewhat more deeply than is the case in Bos primigenius (see Tables XIV and XV).

Pilgrim's diagnosis of Bos acutifrons says, "A Bos of large size; frontals with a prominent median longitudinal ridge (not present in the variety planifrons), length exceeding the width at orbits by about one third. Skull slightly less contracted between horn-cores and the orbits. Horn-cores of enormous size, directed almost entirely outward and without any forward curvature, set very obliquely on the forehead; cross-section at base ovate; occipital very high and narrow, occipital crest either subtriangular or (in planifrons) forming a wide arch in both varieties the summit reaching almost to the top of the upper border of the horn-cores; supra-cristal portion of occiput shallow and rounded off to join the frontals and only slightly overhanging the infra-cristal portion; temporal fossae narrow and slightly indenting the occipital far below the summit of the occipital crest."

Pilgrim then points out certain Bovine affinities exhibited by Bos acutifrons and Bos planifrons. In mentioning these we shall also include the kouprey.

1. "The excess of length over breadth in the frontal." This is not as marked in the kouprey but nevertheless it is characteristic.

2. "The acute angle formed by the planes of the frontal and the occiput." This also applies to the kouprey.

3. "The great reduction of the parietal and supraoccipital regions, as shown by the small interval between the occipital crest and the top of the horn-cores." In the kouprey a reduction has taken place in this region and the supracristal region is curved as in Bubalus. The parietal is not as reduced as in most of the Bos.

4. "The convex curve of the upper border of the horn-cores." There is no such convex curve in the kouprey.

5. "The narrow posterior openings of the temporal fossae and their relatively long distance apart." In this the kouprey resembles the condition in Bibos with wide posterior openings terminating relatively close together on the occiput. Pilgrim points out that Bos acutifrons and Bos planifrons approximate to Bibos in:

(1) "The cross-section of the horn-cores which is flattened from back to front and shows a vestigial keel above." The kouprey is flattened transversely showing a vestigial keel behind and is therefore unlike Bibos in this respect.

(2) "The rounding off of the junction between the occiput and the frontal." In the kouprey this rounding off is much more gradual and at less of an acute angle than in Bibos.

Pilgrim goes on to point out that "certain features in both Bos acutifrons and Bos planifrons are evidence of a stage more primitive than in any known species of Bos or Bibos." This is of special interest for us in comparing the kouprey.

1. "The cross-section of the horn-cores which is identical at any rate in the case of the left horn-cores in Bos acutifrons and Bos planifrons is not merely ovate but is asymmetric on the back and front sides, the former being much more strongly convex than the latter so that the cross-section may be termed subtriangular and is strictly analogous to the condition in Bubalus, except that neither of the two keels so indicated, i.e. the postero-external and the antero-internal (using the same terminology as was employed in the case of the Boselaphinae) is more than vestigial."

The cross-section of the horn-core in the kouprey is even more sub-triangular, only the postero-external primary keel is very evident.

2. "The occipital is characterized by its small transverse diameter relative to its height. In Bibos the occipital is relatively broader than in Bos but in the Siwalik species the occipital is higher and narrower than in either Bos namadicus or Bos primigenius. That this is a primitive character is proved by its existence in Bovine forms such as Proleptobos and Proamphibos as compared with their descendants."

In the kouprey the infra-cristal part of the occipital is relatively higher and narrower than it is in Bos acutifrons and much more so than in Bibos. It has the same general subtriangular shape as in acutifrons with a marked occipital crest. Pilgrim continues, "There seems to me no reason why the Siwalik form should not be regarded as the direct ancestor of Bos namadicus and Bos primigenius, but no idea that it is equally ancestral to Bibos could bear investigation for a moment in the light of the resemblances to Bos listed above, most of which are progressive and not primitive features."

We have shown by our mention of the kouprey under various points of comparison that it resembles Bos acutifrons in those characters that make it more primitive than and ancestral to Bos namadicus and Bos primigenius, but at the same time this resemblance does not apply to Pilgrim's progressive characters that would prevent its being ancestral to Bibos. In fact there are many morphological features that show a closer resemblance between the kouprey and Bos acutifrons than between the kouprey and any hitherto described forms.

Pilgrim considers planifrons a variety of Bos acutifrons differing from it:

1. "In the wider summit of the arch formed by the occipital crest." In this the kouprey is more like Bos acutifrons.

2. "In the absence of the strong median longitudinal ridge on the frontal." In this the kouprey is like Bos planifrons except that there is a faint trace of the upper part of such a ridge described elsewhere in this paper.

3. "The horn-cores are directed more backward." In this the kouprey greatly exceeds either form but it is more like Bos planifrons than Bos acutifrons. In commenting on the difference in size Pilgrim says it is less than one would expect to be the case if one were the female of the other. He also says that the shape of the occipital crest in the Bovinae is variable with age, sex or individual and probably the occiput of these two forms differed little from Poephagus grunniens except for the greater height of the occipital crest. He does not separate Bos planifrons specifically from Bos acutifrons.

The comparative measurements can be seen from Table XIV compiled from Lydekker's measurements. Cossar Ewart (1911) points out that a Newstead skull which he figures shows a deeply notched occiput and in this bears a much more marked resemblance to Bos acutifrons of the Pliocene Siwaliks than to such extinct forms as Bos primigenius and Bos namadicus. He concludes that some of the cattle in the South of Scotland during the Roman occupation were descended from an Indian race allied to Bos acutifrons.

Pilgrim says, "Bison sivalensis agrees perfectly with the living and Pleistocene species of Bison. Those characters in which it differs we are justified in regarding as primitive such as one might find in a Pliocene species which was ancestral to the modern bison. These are: (1) the narrower occipital, (2) the greater indentation of the supraoccipital by the temporal fossae; the compression of the horn-cores which gives some indication of an anterior and an inner keel.

"That a narrow occipital is primitive is almost self-evident, if we believe that the Bovinae are descended from antelopes. With regard to (2) we may suppose that the distance apart of the hinder ends of the temporal fossae remained as in the antelopes and did not keep pace with the general widening of the skull. At any rate in the two most primitive Bovinae known, Proleptobos and Proamphibos, not only is the occiput narrower but the temporal fossae are closer together than in their lineal descendants.

"Bibos exhibits the same difference from Bos, in which genus, equally with Bison, the spacing of the temporal fossae has increased in conformity with the general occipital widening."

The points that Pilgrim indicates about the occiput of Bison sivalensis are even more applicable to that of the kouprey as may be seen by the following measurements from Lydekker.

	Bison sivalensis	Novibos
Width of palate at M^3	87.5	83
Interval between external angles of occipital condyles	100	106
Width of occiput between temporal fossae	103	75
Height from lower border of foramen magnum to occipital crest	124	142
Antero-posterior diameter of horn-core	87.5	10.4
Transverse diameter of horn-core	67	62

As to Pilgrim's comment on the compression of the horn-cores, if we agree that the general development of the Bovine horn-core follows his outline, "that is an internal expansion with the production of an inner keel in addition to the two existing primary keels, followed by a gradual obliteration of two or more keels resulting in a more or less rounded cross-section, then the cross-section of the horn-core in Bison sivalensis must be regarded as representing a stage of development where the inward expansion remains and there are relics of both anterior and internal keels although the posterior keel has vanished."

The horn-core of the kouprey represents an even earlier stage which retains the posterior keel and a suggestion of the primary anterior keel without as yet producing an inner keel.

Pilgrim considers that "any likeness which Bison sivalensis bears to Poephagus grunniens it owes to the fact that it lies nearer than other species of Bison to the point from which Bison, Poephagus and Bibos diverged and which must in my opinion have been previous to the Upper Pliocene (Villafranchian). Between this common hypothetical ancestor and the living yak no link appears yet to have been discovered." He substantiates his statement by pointing out that "Poephagus differs from Bison in almost every essential character except the possession of 14 ribs, a short premaxilla, a low occiput and two narrow temporal fossae. Bos also has a narrow temporal fossa and a low occiput. Bos shares with the yak a relatively long face and nasal, while the deflection of the parietal into the plane of the occipital and the narrow bulla are characteristic both of Bos and Bibos."

Pilgrim's placing of the point of divergence of Bison, Poephagus and Bibos previous to the Upper Pliocene also means that the kouprey which combines characters found in all three of these forms in their primitive manifestations, must have emerged previous to the Upper Pliocene and probably in the Middle Pliocene. The kouprey appears to be a closer direct descendant of the hypothetical common ancestor of the living Bovinae than any other described living form.

To sum up our comparison with fossil Bovidae based on Pilgrim's recent monograph: the kouprey shows a more primitive condition than the living Taurina and than many of the fossil Bovinae such as Bos namadicus, Bos primigenius, and Bos acutifrons, and Bison sivalensis. It has certain characteristics that suggest a closer resemblance to Bos acutifrons than to the other above-mentioned forms.

The primitive features of the kouprey skull are based on a comparison with Pilgrim's analysis of primitive and progressive features in (1) the cross-section of the horn-core, (2) torsion of the horn-core, (3) situation of the horn-cores, (4) divergence of the horn-cores, (5) the frontal bone, (6) the supraorbital foramina and pits, (7) the lachrymal bone and fossa, (8) the premaxilla, (9) the parietal bone, (10) the temporal crests of the parietal, (11) the occipital bones, (12) the basicranium, (13) the palatal bone and the vomer, (14) the dentition, (15) the metapodials. This comparison results in the necessity of placing the kouprey skull in Pilgrim's phylogeny at some point preceding the branching off of the living genera of Bison, Poephagus, Bibos and Bos but not as far back as Proleptobos and Leptobos. There are a number of points of resemblance between the kouprey and Proleptobos and Leptobos. These fossil forms are however even more primitive, especially in the occipital region. It is quite probable that Novibos separated from the Proleptobos-Taurina stem at some time in the middle or late middle Pliocene (see chart fig. 9 p. 494).

TABLE XIV

INDIAN FOSSIL BOVIDS (measurements from Lydekker, 1878) AND KOUPREY

Measurements (mm.)	Bos namadicus	Small Bos namadicus	Bos planifrons	Frontlet Bos planifrons	Bos acutifrons	Bos platyrhinus	Bison sivalensis	Novibos sauveli
Length from foramen magnum to diastema	318							280
Length from intercornual ridge to apex of nasals	330	241	259		330			200
Height from lower border of foramen magnum to occipital crest	147.3	114.5			147.3		122	140
Interval between occipital crest and intercornual ridge	63.5	50.8	50.8	45.7	45.7			52?
Width at inferior border of orbits	218.2					211		207
Width at superior border of orbits	254	147.3	236		279		251	202
Width below orbits	188	160					142	129
Width above orbits	206	170	223		221		221	170
Width of occiput through petrosals	262				287			227
Height from surface of palate to frontals	173							
Greatest width of occipital crest							17.4	160
Interval between outer angles of condyles	136	96.5					99	106
Length of temporal fossa	170	134.5			147.3		136	135
Length of intercornual ridge	235	157.5	178	183	127			118
Height from palate to tip of nasals	81.3							133
Interval between orbit and base of horn-core	134.5	86.3			105			75
Transverse diameter of left orbit	68.5	55.8			60.9		55.7	58
Antero-posterior diameter of left orbit	78.7	63.5			66		55.7	58
Length of line of five molars	150							
Greatest width of palate	96.5	96.5						83
Interval between outer surfaces of second molars	147.3	136				152.5	(M3) 142.3	125
Length of left horn-core, upper surface	989							
Length of left horn-core, lower surface	812							553
Distance from inferior border of foramen magnum to last molar							160	187
Length of last molar							35.6	25.5
Length of two molars							68.5	48.5
Anterior-posterior diameter of base of horn-core	(left) 96.5	63.5	(right) 114.5		152.5		86.3	104

Measurements (mm.)	Bos namadicus	Small Bos namadicus	Bos planifrons	Frontlet Bos planifrons	Bos acutifrons	Bos platyrhinus	Bison sivalensis	Novibos sauveli
Circumference of above	317	152.5	368	368	432		254	268
Transverse diameter of above	86.3	60.9			101.5 (broken ends)		66	62
Interval between tips of horn-cores	913				2,210			730?
Interval between supra-orbital foramina	144.7							99
Width of nasals at widest part	68.5					88.8		66
Width of occiput between temporal fossae		119.5	170	178	157.5		105	75
Width of palate at second molars						91.4		
Width of palate at third molars							86.3	83
Length of six molars		139.7				152.5		118
Length of last molar		35.5						25.5
Length of nasals						198		202
Height from posterior extremity of palate to frontals						198		144
Length from posterior extremity of palate to muzzle						305		302
Width of maxillae at diastema						114.5		98
Width of frontals between upper angles of horn-cores							209	94
Same between lower angles of horn-cores							262	190
Height from palate to centre of nasals						134.5		103

TABLE XV

MEASUREMENTS (in millimeters) OF FRONTLETS AND HORN-CORES OF URUS IN THE BRITISH MUSEUM (After Reynolds)

	Herne Bay	Ilford, 45430.	Ilford, 45427.	Ilford, 45437.	Clacton, 2342.	Clacton, (left) M. 15624.	Kouprey M.C.Z. 38108.
Width between bases of horn-cores	210.	315.	305.	—	240.	—	ant. 190 post. 94
Circumference of horn-core at base	460.	447.	463.	448.	477.	485.	268.
Length along outer curvature	963.	865.	960.	797.	770.	870.	553.
Span of horn-cores from tip to tip	837.	624.	854.	—	954.	—	730?

TABLE XVI

MEASUREMENTS (in millimeters) OF SKULLS OF BOS PRIMIGENIUS (AFTER REYNOLDS)

	Ilford (Brady Coll., Brit. Mus., no. 50086).	Athol (Brit. Mus., no. 2245).	Preston (Brit. Mus., no. 8913).	Chingford (Brit. Mus., no. 10353).	Kirkeudbright (Brit. Mus., no. 36405).	Ham Moor near Newbury (Newbury Mus.).	Kouprey M.C.Z. 38108
1. Length from occipital ridge to upper end of nasal	355.	375.	342.	352.	310.	364.	200.
2. Length from occipital ridge to lower end of nasal	625.	630.	608.	583.	553.	602.	402.
3. Space between roots of horn-cores measured along occipital ridge	215.	263.	218.	198.	190.	—	118.
4. Width of skull across upper edge of orbits	335.	325.	324.	315.	233.	310.	175?
5. Circumference of base of horn-core	445.	351.	378.	380.	348.	392.	268.
6. Length of horn-core following outer curvature	767.	724.	770.	732.	551.	596.	553.
7. Span from tip to tip of horn-cores	791.	1028.	608.	613.	521.	800.	730.?

POSSIBLE BEARING ON ORIGIN OF DOMESTICATED CATTLE

This subject associated with research on fossil cattle has been widely discussed and additional information, that might have a bearing on it, discovered from the time of Pallas' paper in 1768 to Reynold's and Pilgrim's papers of 1939, of which the last two have been extensively used in this report. References to the more important papers are given in the bibliography. Early in the 19th century the accounts of Cuvier, Bojanus and von Meyer stand out conspicuously after which, in the middle of the century, came Owen in England, Nilsson in Sweden, and Rütimeyer in Switzerland. These are followed in the second half of the century by Darwin, Dawkins, Brandt, Clark, Hughes, Forrest, Lydekker, Middendorf, Nehring, Pomel, Werner, Wilckens, Frantzius, Pagenstecker, Adametz and Keller. During the present century the writings of Johnston, Duerst, Auerbach, Mertens, Siegfried, Fiedler, Staudinger, La Baume, Boule, Pohlig, Ritchie, von Luthner, de la Fosse, Gromova, Rostafinski, Maxim, Morse, Reynolds, Schwarz, and Pilgrim are of importance.

In Morse's excellent discussion of the conflicting views on the "Ancestry of Domesticated Cattle" he sums up the most important opinions.

There seems to be increasing evidence from our archaeological knowledge of the early inhabitants of Central Asia that long-horned cattle were probably domesticated there at an earlier date than in central or western Europe or Africa. Through the courtesy of Mr. Lauriston Ward of the Peabody Museum I have reviewed a recent unpublished survey study made by Sherwood Washburn as to the actual evidence of the presence of domestic animals in the early Near-Eastern sites especially Anau, Fayum, Merinde, and Nineveh. Washburn concludes that "cattle, sheep, goat, pig, and dog form an early group which goes back as far as the earliest levels considered in this report," and adds that "these conclusions seem reasonable in the light of present evidence but are by no means proved." Domestic cattle were probably brought westward with the early Asiatic invaders in Neolithic times and gave rise to the various domesticated races of Bos longifrons such as the Celtic shorthorn. These cattle were more intimately related to the Asiatic zebu according to Ewart than to the wild urus, Bos primigenius of Europe. Wilckens (1880) felt that the Swiss lake-dwellers must have received their marsh cow after its domestication in Asia for the animal is too small to stem from "primigenius" and besides its remains have not been found wild. He also says that the only alternative is to derive the marsh cow from Bibos. The zebu is the nearest to Bibos of any of the taurus group and is not found wild except when it has escaped from captivity. Troltsch (1902) regards the domesticated "primigenius" forms as steppe cattle which came from Asia to Europe about 3000 B.C. In Keller's classification (1902 and 1905) he has two basic forms, Bos primigenius of Europe and Bos sondaicus, whose home he says was in Java. In the latter group he includes "the Asiatic and African Zebu, Old Egyptian Longhorn, Algerian, marsh cow of the lake dwellers, Albanian, Sardinian, Spanish, Polish, Channel Islands, Hornless Fjell, and Brown Swiss breeds."

According to Duerst (1908) the ancient Sumero-Accadian ideogram for a bull =  and wild bull is the same with the sign for mountain enclosed making it a "bull of the mountains" = . About 8000 B.C., Duerst notes the appearance of a domesticated long-horned ox which he regarded as a domesticated form of namadicus identical with what he calls Bos taurus macroceros of Egypt which was spread at a still later period by tribal migrations.

The other domesticated animals of western Asia are much the same as those of Egypt. The zebu was domesticated probably as early as 4000 B.C. and spread from Asia to Africa, so that the ancient distribution was practically as it is today. Aristotle, Pliny and Oppian knew of the zebu in Syria, and it may have gradually changed into the steppe breed.

"Hartmann was convinced there were three breeds and made his division according to the shape of horns—the lyre-form, the half-moon, and those in which the horns pointed away from each other."

Adametz says (1894) since no fossil remains of zebus have been found in central and southern Africa they are of Asiatic origin with the bantin as their ancestor.

Many of these opinions agree in tracing the ancestry of Bos longifrons through Egypt back to Asia from a form ancestral to the zebus. Adametz, Keller and some others have selected the bantin (Bibos banteng) as the probable living representative of the ancestral form. There is no doubt that the bantin is more primitive than the modern zebu in the temporal fossae opening more widely on to the occiput. The horns have lost all traces of keels and are comparatively small with a simple circular curve.

Morse says: "Summing up the most important of these heterogeneous opinions, we find that Cuvier, Werner, Middendorf, and Nehring believe in a monophyletic origin of European cattle, and that Bos primigenius was the only ancestor and its home in Europe. Adametz and Dürst believe in a diphyletic origin, the ancestors being primigenius and longifrons and the homes of both in Asia; while Rütimeyer, Frantzius, Pagenstecker, Wilckens, Keller, and Ewart, though believing in a diphyletic or polyphyletic origin, think that at least the home of longifrons was in Africa or Asia."

Morse has wisely stated, "All species of Bos which lived in the Pliocene and Pleistocene are now extinct, although the blood of several forms of Bos primigenius, Bos namadicus, and Bos priscus may still flow in the veins of our domesticated cattle. To these species and the representatives of the genus during the Recent or Alluvial periods we must look for the genealogy of our cattle. On the other hand, probably all of the species of the Recent period have played a part in the history of cattle raising."

With this we entirely agree. It leads us to examine the kouprey with a view to determining what part its blood might have played at a remote or a recent period in the development of domesticated cattle. In some ways the kouprey makes a more fitting living relic of an ancestral form than the bantin. It resembles the zebu and early long-horned cattle more closely by its long tail than does the Bibos group. Reynolds (1939) speaks of a long tail as a primitive feature whereas Schwarz (1937) calls it a progressive one. If we follow Williston's law we are inclined to agree with Reynolds in this point. In the kouprey skull such primitive features as the short tooth row, quadrate teeth, narrow skull with a sloping almost Bubaline supracristal occipital, the temporal fossae greatly indenting the occiput, the backwardly directed horns with primitively shaped horn-cores retaining a primary posterior keel, the ethmoid vacuities, and the general lyre-shaped curve of the horns which are not set as far back on the skull as in the more progressive forms, all help to qualify this animal as a probable living descendant of an ancestral form not only for the yak and bison but also for Bibos and Bos.

In his 1902 paper Keller draws an able comparison between a narrow-skulled female bantin and the zebu mentioning especially the flattened eye sockets, broad and shallow temporal groove, a narrow ethmoid vacuity, and the premaxilla that doesn't contact the nasals. He says that domestication has shortened the length of the face and increased the length of the forehead. He hails the discovery of an Egyptian stone carving in the Louvre that may date from the first dynasty which shows in the direction of the horns, the exceptional width of the forehead and the short muzzle, all the characteristics of an old bantin steer. He may be right, but a close relative of the kouprey would satisfy most of his comparison requirements and at the same time would have an occiput more like that of the zebu, a skull more primitive rather than progressive in such points as the lengthening of the frontal, a dewlap instead of having to develop one, a long tail rather than forcing the zebu to evolve a long tail from the much shorter-tailed bantin; less divergent horn-cores and a profile of shape of skull that constantly reminds one of sculptured figures of Egyptian, Assyrian and Iranian cattle rather than just one or two that might suggest an atavistic throwback to the ancestral form. If Keller had known of the existence of the kouprey I am confident that he would have found in it, in spite of its primitive features, a closer possible ancestor of the zebu than the bantin. A detailed comparison in the first part of this paper has shown the many resemblances of the kouprey to the bantin as well as the progressive features such as the narrow nasal width and wide foramen magnum, and large orbits which the bantin does not share with the zebu.

Is it not conceivable that the kouprey, about 6000 B.C., may have had a comparatively localised range, and also not being of gigantic size like Bos namadicus and Bos primigenius it could more easily have been brought into domestication or interbred with captive smaller forms of Bos namadicus and Bos acutifrons.

There seems to be a desire among some of the students of this subject to derive the smaller cattle of the Bos longifrons type from such gigantic species as Bos primigenius and Bos namadicus. The records of palaontology show us that while giants have arisen in many groups of mammals, in such stages they have advanced beyond their period of plasticity and soon become extinct. One needs only to examine the skull of a large Bos primigenius or Bos namadicus to see how progressive it is in many important features as compared with the living Bibos or Novibos group.

The races of domesticated cattle related to the zebu may contain some blood of the giant wild Bos of Europe and Asia, but their skulls indicate that a more primitive form has played a part in their ancestry. We know that in no instance have fossils shown us that giant forms of one stock can be modified into a later giant of an obviously different type. There is no indication that domestication can bring about such a result, and therefore we should expect the skulls of bulls of the largest breeds of domestic cattle to differ markedly from those of Bos primigenius and Bos namadicus which they do. It is more probable that domestic cattle were descended from smaller ancestral forms that may have shared some of the blood and inheritance of the gigantic fossil forms.

It is of interest to note which of the basic types of Indian cattle would most nearly resemble the present wild kouprey. From Col. Sir Arthur Oliver's "Brief Survey of Some Important Breeds of Cattle in India" (1938) it would appear that the closest resemblance would be to the Mysore-type cattle of the South. He says "This is one of the types that has been longest in India and in which there is the least likelihood of heavy admixture of foreign blood. In this breed the head is comparatively long with a narrow face and nostrils. The forehead is usually prominent and the horns emerge from the top of the poll fairly close together extending upward and backward rather than outward and upward as in other breeds. The Amritmahal Breed is one of the best known from Mysore. They are active cattle bred for draught purposes and famous for their power of endurance. They have small pointed ears, a well-developed dewlap and a hump placed in front of the withers. They have a very small sheath and close fitting skin with a fine tail of moderate length with a black whisp reaching only a little below the hock. Their general coloration is gray with dark head, neck, hump and quarters and light gray or white markings about the face and dewlap. Their frame is compact with straight limbs of fine quality."

The Killari Breed as described by Colonel Oliver is a derivative of the Amrit-mahal and may have a strain of the gray-white cattle of the north in its composition. They have a gray-white coloration, a comparatively long body with well-arched ribs, a massive head but not such a prominent forehead as the usual Mysore type. The long sweeping horns emerge near each other at the top of the

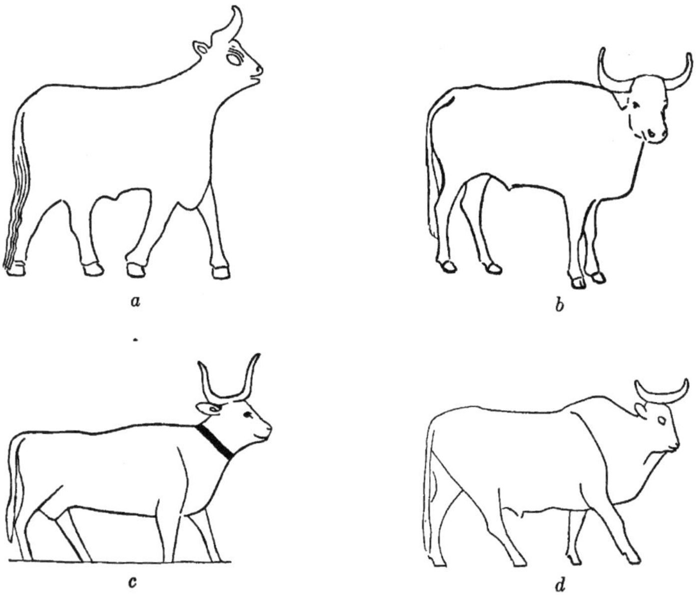

Fig. 10. Common forms of cattle in Egyptian and Mesapotamian art compared with a 4-year old Kouprey drawn from a photograph. These outline sketches show the characteristic long tail, indications of a dewlap, and shapes of body and horn. See p. 512.

a. Outline of Mesapotamian bull sculptured in relief before 1500 B.C. (Zervos p. 66).
b. Outline of Kouprey bull from photograph in Vincennes Zoo 1939 (*Mammalia* Dec. 1939).
c. Outline of long-horned Egyptian cattle about 3500 B.C. (Lydekker 1912 p. 135).
d. Outline of Egyptian bull about 1500 B.C. (Weigall p. 135).

poll in an upward direction. They have a narrow face, a well-developed hump in front of the withers, a voluminous dewlap, and small sheath. The comparatively short tail with a black switch reaches only a short distance below the hock.

Probably in these two breeds we have the closest resemblance to the kouprey among the living races of Bos indicus. The large dewlap, small ears, slender limbs, elongated body, moderate length of tail, gray-colored skin varying from a light shade to almost black, the fine limb bones, fine skin close to the body and the long narrow skull with horns emerging from the poll; all suggest characteristics shared by the kouprey. The development of the hump is in all probability a product of domestication.

I do not wish to suggest a relationship of these cattle with the kouprey, but only to draw attention to the fact that they have certain external resemblances that may be of interest in a further investigation of this subject. Unfortunately I was unable to obtain in this country a skull of one of these Mysore cattle for comparison with that of the kouprey. This is not one of the breeds that we have imported for use in America.

a *b* *c*

Fig. 11. Outlines of heads of bulls showing thickening at base of horns and suggestion of notched nostril. See p. 513.
a. Egyptian 663–342 B.C. (Weigall p. 329).
b. Babylonian 605–561 B.C. (Koldewey p. 43).
c. Boetian 700–600 B.C. (Greek Vase Boston Art Mus.).

As Reynolds points out, cattle are one of the most highly specialized groups of ruminants. With the increase in size and weight we should expect the development of short and massive metapodials. Those of the kouprey are slender and quite similar to those of the bantin.

In addition to the primitive skull characters of the kouprey which help to qualify it as a contemporary ancestor there is the long tail and the white stripe on the back which is a characteristic wild Urus (Bos primigenius) marking.

If we compare the outline of the 4-year old kouprey in the Paris Zoo (fig. 10 b.) with figures of long-horned cattle so frequently seen in early Assyrian and Egyptian sculpture (fig. 10 a.c.d.) the similarity of the shape of the body of the animals is at once striking.

The difficulty of drawing any conclusions with comparisons of cattle as figured in early sculpture and pottery design in Egypt, Greece or the Near East is that the artists were required to follow certain conventions that distorted the natural appearance of the ox in question. However, it is interesting to note that all the figures of cattle showed long-tailed forms, that some showed the development of a dewlap without having a hump (fig. 10a.) and that the profile of the base of the horns made them appear quite thick close to the base. The kouprey has a

a *b*

Fig. 12.

a. Outline of skull of Sacred Apis about 3500 B.C. Note the general shape of the skull and approximation of the bases of the horn cores. (Keller, 1909, p. 331).

b. Outline of a conventionalized bull's face from Greek pottery about 550 B.C. Note the notched nostrils, elongated face, and small ears. (Morin p. 127).

marked notch in the outline of its nostril. The shape of cattle nostrils depends to some extent on the mood of the animal, and most of them show some degree of notching. I have outlined from ancient sculptures the following characteristic heads of a Greek, an Egyptian, and a Mesopotamian bull, two of which show nostril notches and two the conventional horn profile widened close to the skull. (fig. 11 a.b.c.) The front face outline (fig. 12 b.) is Boetian 6th century B.C. and indicates a long narrow-shaped skull as well as notched nostrils.

These long skulls recall Marriete's discovery in 1851 of skulls of the sacred Apis in the Serapeum of Memphis. This (fig. 12a) is the outline of one taken from a photograph of a specimen preserved in the Egyptian Museum in Giza. These bulls were chosen from the stately powerfully built race of long-horns without humps praised by Aristotle and now extinct. The skulls were characterised by a rapidly sloping forehead, slightly protruding eye sockets, horn-cores approximating one another and a relatively well-chiseled face. One is also struck by the delicate finely chiseled muzzle in the old reliefs.

There are constant suggestions of parallelisms showing the possible effect of an Asiatic bovine heritage in which the kouprey might well have been a remote ancestor. The domesticated cattle are a highly variable group. Among the zebus of Asia and Africa one finds giant forms and pygmies, large-humped and no-humped, long-horned or polled, narrow or broad-skulled. No such wide variation is known to exist in the living genera of wild Asiatic cattle, although perhaps the most variable forms are individual, sexual and racial differences in the bantin of the Bibos group. We have no information as yet about the variation in Novibos for even the female form is undescribed.

The locality where the kouprey is at present found is not far from Northern India which has been considered the cradle of the Asiatic Bovids.

In speculating on the possibility of the kouprey being merely a hybrid between some Bibos and a domestic zebu, it is inconceivable that the resulting skull could develop such primitive features as those that characterize the kouprey. Even if some buffalo blood were introduced into the combination, the resulting "cataloe" would be very different from the kouprey as evidenced by existing cattle-and-buffalo crosses.

Further study may well reveal that in the kouprey we have a more probable living relative of the remote ancestor of the first domesticated long-horned cattle of Anau and Egypt than we have in the living Bibos group. Similarly we may have a contemporary ancestor of the extinct but Recent and Pleistocene forms, Bos primigenius of Europe, Bos namadicus and Bos acutifrons of Asia.

The Anoa still living in Celebes is a primitive survivor of an ancient Bubaline which has retained its small size. The kouprey is in a somewhat similar position (see chart p. 494) being a primitive survivor of an ancient Taurine, which has, however, the characteristic cattle feature of large size and well-developed horns. It seems strange that it should have survived in such a primitive form without an isolated habitat such as living on an island. It is the more remarkable when we realize how easily even different genera of cattle are hybridized in domestication. Nevertheless there is a stretch of several hundred miles in Northern Cambodia and Southwestern Laos that is still comparatively little known to the white man, while the scattered natives in the entire region have probably known of the existence of the kouprey for generations.

In order to show how little we really do know about the kouprey I have listed and discussed the available documents dealing with this subject in the Historical Appendix.

We can only hope that in the coming years the French Indo-China Governmental authorities will make every effort to preserve the kouprey from trophy hunters and native poaching, at the same time making it possible to carry on further investigation with regard to the life history and anatomy of "Novibos," an additional wonder, that like the unforgettable temples of Angkor, has long remained buried in the tropical forests of that beautiful country.

HISTORICAL APPENDIX

Our scientific knowledge of the kouprey is so recent that it seems advisable to record it chronologically with a brief bibliography at the end of this section. The natives of certain parts of Northern Cambodia and a few of the white hunters in the neighborhood of Saigon have undoubtedly heard for many years of the existence of a wild ox in parts of Northern Cambodia that differed from the well known bantin and the gaur. In August, 1929 after several months' collecting small mammals and birds for the Field Museum in Northern Tonkin and Laos on the Kelley-Roosevelts' Expedition I joined Theodore Roosevelt in Saigon. He had just completed a rainy-season hunt with Defosse and his son Louis, during which he procured skins and skeletons of gaur and bantin as well as various deer for that museum. [1]I distinctly remember Defosse mentioning in conversation an interesting little-known wild ox that he had heard of. The logical suggestion was made that it might be a bantin-gaur hybrid or some strain of native cattle that had gone wild. The following year Defosse (1930) sent me his privately printed circular to interest hunting parties in which he definitely refers to the kouprey.

As far as I know this is the first printed reference to the kouprey and is referred to in Professor Urbain's original description. Urbain also mentions the publication by Dr. R. Vittoz, Veterinary Inspector of Cochin China (1933). This author at the time writes of the kouprey in an article on the raising and exploitation of Bovines of the southern Indo-Chinese in Cochin China (1933, p. 956):

"Le boeuf gris—(cambodgien: Kou prey)—est un animal nettement différent du Banteng par sa grande stature et son pelage entièrement gris chez les jeunes

[1]Coolidge and Roosevelt. 1933. Three Kingdoms of Indo-China. T. Y. Crowell Co.

et les femelles et d'un beau noir mat chez les vieux taureaux. Le front des animaux de cette espèce est dépourvu de dépression, le chignon est étroit non carapaçonné et porte des cornes cylindriques largement écartées, recourbées en avant chez le taureau et en forme de lyre chez la vache. Comme le Banteng, le boeuf gris possède un garrot prolongé en arrière sur la région dorsale et porte quatre balzanes.

"Cet animal très rare, n'existe plus actuellement que dans certaines régions de forêt-clairière peu fréquentées du Nord Cambodge.

"Le Kou prey signalé par le Dr. M. Dufossé dans son livre intitulé "Chasse et Tourisme au Cambodge et dans le Sud-Indochine" a été rencontré à plusieurs reprises par le Docteur-Vétérinaire Sauvel, Chef du secteur du Nord-Cambodge, qui a effectué des mensurations de cet animal et rassemblé de nombreux renseignments inédits sur l'habitat du Kou prey et ses rapports avec les autres bovidés sauvages."

Mr. Donald Carter in the Mammal Department of the American Museum of Natural History has interested himself in the kouprey for some years and has kindly made available to me the information that he has gathered. The most important is a letter dated March 7, 1936, which was written to him by Mr. J. F. Richmond of the Standard-Vacuum Oil Co. from Hong Kong. He says,

"In January this year I went hunting in Indo-China. Monsieur Vincent Pietri of Saigon acted as my guide and excited my curiosity considerably by talking about a strange, unknown beast.

"The animal is known to the natives and is called 'KOU-PREY' by the Mois. The French spelling of the word is derived by me from Pietri who maintains that there is no other name.

"Pietri personally has killed a 'kou-prey.' I forget the date, if it was mentioned, but it was during the last three years. Pietri took the head for mounting but the skin was spoiled and he now has only the horns. At the time the animal was strange to Pietri but he thought it was new to him only and did not think it was actually a new and unknown beast to everyone. This he found out later from the continual questions about the horns asked by hunters, collectors and others with whom he comes in contact as a taxidermist and professional guide.

"Pietri's kou-prey was killed in Cambodia about 200 kilometers north of Saigon near or not far from the Saigon-Kratie highway. It was discovered in high grass in open woods, and was a species of wild cattle or buffalo being a bull about 5 feet high at the shoulder and entirely black in color (except that I am not sure whether Pietri mentioned that it had light stockings like a gaur). The animal was a solitary.

"I attach a photo taken by me from a snapshot taken by Pietri when bringing the head into camp. The head is tied to a pole supported on the shoulders of coolies who do not appear in the picture. I hope this picture can assist you in identifying the animal or in concluding that it is really a new species. Please note the frayed condition of the tips of the wide, heavy horns which I have myself seen in Pietri's Saigon shop. Their spread is about 42 inches."

Examination of the photograph of Pietri's snapshot of his first specimen indicates that the head was in all probability that of a kouprey which according to Mr. Richmond's information was shot between March, 1933 and March, 1936.

The frontlet which we have used for comparison (Plate III, fig. 3), loaned to us through the courtesy of the American Museum of Natural History, was shot by Mr. Ezra B. Cornell of Rumford, Rhode Island, in February, 1939. The following are extracts from two letters of his correspondence with Mr. Carter concerning the specimen, written by Mr. Cornell:

(March 1, 1939) "Just prior to leaving for the States the writer was on a big game hunt in northern Cambodia, French Indo-China, and on the last day of the expedition shot one of the rare animals mentioned in Mr. Richmond's letter to you. The animal is decidedly different from any other of the wild cattle family, such as the gaur, banteng, etc., which I have shot in Indo-China. There were two together, both of them having frayed tips on the horns and the heads being identical with the photograph which Mr. Richmond sent you. Unfortunately the bull I shot ran for about 12 kilometers before dropping and the natives who brought the head in had lost the scalp. However, the horns are in my home in Rhode Island where the attached photograph was recently taken with the thought you might be interested;"

(and March 9th, 1939) "Thank you for your letter of March 7th which is very interesting. I shall be pleased to forward the trophy mentioned to you for inspection.

"In answer to your question, the narrowness of the forehead between the horns seems to be distinctive of this type of animal, from what I can gather in talking to Mr. Pietri. Also the head being coal black gives it a distinctly different appearance from the gaur which has a lighter colored forehead.

"The animal I shot, as previously mentioned, ran for a considerable distance and the coolies brought in the head later, therefore I cannot give you as accurate a description as I would like. However, both these animals, although first seen at a distance of about 100 yards, passed by me after the first shot at a distance of about 20 yards, which gave me an opportunity for a very good look (their ability to absorb lead is amazing since both were fatally wounded). From this distance both animals appeared coal black, there being no sign of white stockings, although these markings could have been obliterated as they were both running through a burned country. Would estimate their height at between 5 and 6 feet and the body appeared to me to be shaped like the banteng. One thing was distinctly noticeable, i.e. that both animals carried their heads high when running which brought out the upsidedown tassel effect near the tips of the horns."

Dr. William K. Gregory and Mr. Harry Raven of the American Museum both reported seeing two mounted heads of the kouprey exhibited in the Indo-China Pavilion at the San Francisco World's Fair. These were easily recognized by the frayed tips of their horns.

Earlier in this paper we have discussed Duerst's illustration published in 1926 showing the Paris museum skull of a "Boeuf des Stiengs" from Cambodia. This is probably a kouprey x zebu or kouprey x native cattle hybrid.

We are indebted to M. Achille Urbain for the first scientific description of the kouprey (1937) in which he names it "Bos (Bibos) Sauveli" in honor of Veterinary Doctor Sauvel who he reports possesses remarkable trophies of this animal. Dr. Sauvel succeeded in capturing a young live male kouprey that is now in the Zoological Park at Vincennes. He let M. Urbain examine a male that he had just killed near the village of Tchep in Northern Cambodia. With the original description is a photographic plate showing a mounted kouprey head (probably the property of Dr. Sauvel), a young adult bull without frayed tips lying where he was shot, and the live bull calf which is now the holotype at the Vincennes Zoo.

The original description of two printed pages referred to in more detail elsewhere in this paper mentions that this very rare animal lives in the savannah forest (forêts-clairières) where few hunters have been able to approach it. The rest of the description details certain external characters in which the kouprey differs from the gaur and bantin with a special mention of the shredding near the tips of the horns of old males. M. Urbain says that Dr. Sauvel will publish elsewhere the exact measurements of this animal and will make a detailed study of the skeleton. His observations will be most welcome in view of the primitive characteristics of the skull described in this paper which distinguish it as a new genus, as well as differences in the skeleton that can only be verified by observations on additional material.

In December, 1937, Professor Urbain's original description was reprinted in *Mammalia.*

In May, 1938 Professor Urbain's description was published in Amsterdam.

In 1939 there appeared a pamphlet entitled "Big-Game Hunting in French Indo-China" written by M. Omer Sarraut for the Golden Gate International Exposition of San Francisco. The author devotes a page and a half to the kouprey in which he says, "This is a curious bovidae of the upper Cambodia and lower Laos districts. It is little known and has only recently been classified."

"The professional hunter Pietri and myself are the first I believe to report its characteristics." In referring to the frayed and jagged horns in old animals he says, "Natives attribute this to the fact that the kouprey uses its horns to knock down anthills in order to consume the ants. . . . This explanation is given with the utmost reserve."

"The cow is smaller than the male and carries lyre-shaped horns."

"The best season for tracking this animal is between January and June."

In the September-December number of *Mammalia* 1939 there was published with photographs an article on the collection of "Asiatic Bovids of the Zoological Parc at Vincennes" by Ach.Urbain, P. Rode and M. A. Pasquier. This article describes a gaur, gayal, bantin, kouprey and two water buffaloes, all living in the park.

The kouprey is described as 3½ years old coming from the region of Tchep in Northern Cambodia having been at the park since April 1937. This is the first living specimen to be brought to Europe. His coat which was gray on arrival at the park has already darkened. The hair is short and fine making his coat velvety. The lower limbs are white but the front of the forelegs is black, as well as the area of the back legs just above the hoofs. The extremities are "très fines" and the feet "très effilés." The tail is very long falling almost to the hock.

The cylindrical horns are light at the base, black at the tip. They are widely spread and curved forward. "Striés" delimiting the outer superficial layer of the horn are visible at their tips. The ears are slender; the dewlap is very marked while the dorsal ridge is slightly prominent. This animal is of lighter and more graceful build than the other Asiatic bovids. Accompanying this article is a page of photographs showing a side view of each of the young Asiatic bovids described in the article. A comparison of the pictures shows clearly the points already mentioned. The body of the kouprey appears to be distinctly more graceful, and seems heavier forward than the gaur or gayal. The bantin shown on the plate is not comparable as the animal is only 14 months old. The kouprey's coat appears to be sleek compared with that of the gaur and the prominently notched nostrils are plainly visible.

In preparing the present report we noted that no holotype had been designated for the new species, therefore as a result of an inquiry by Mr. Greenway, M. Achille Urbain presented a paper (1939) for the Academy of Sciences in which he for the first time designates a holotype for the kouprey. This is the male captured in July, 1936 close to Tchep in North Cambodia and living at the Vincennes Zoological Park (Paris). He calls the animal adult, 4 years old in December, 1939. In briefly noting the characters he mentions the height at the withers as 1 m. 57 mm. The description is essentially the one given in his previous papers already referred to with the additional note, on a point we especially inquired about, that is that the eyes are chestnut colored (they are blue in the adult gaur). The geographic distribution is given as "Cambodia" and the habitat as the "forêts clairières." In the *Bulletin du Museum* (2e s., t. XI, no. 6, 1939) the last mentioned article is republished with the addition of the kouprey photograph from the *Mammalia* 1939 article.

The shooting of our specimen is to be described by M. François Edmond-Blanc in an as yet unpublished article. The fact of the arrival in France of this specimen of a "Kou Proh" was described in a newspaper article in *Le Jour-Echo de Paris* by Count de la Chevasnérie with a front view photograph of the dead animal.

M. François Edmond-Blanc was associated with M. Jean Delacour of Clères and James C. Greenway, Jr., of this Museum on the "VIIe Expedition en Indo-Chine." During the first part of March, 1939 while Greenway was collecting birds in Annam Edmond-Blanc with A. V. Pietri, the Indo-China hunting guide, succeeded in shooting the very fine male kouprey which is herein described. It is reported that the hunt took two weeks with more than 400 kilometers of tramping on foot across the plains and open forests of the Cambodian region of the Phnom with a temperature of 35° C. in the shade. Once the animal was located the final tracking took seven hours. (See map, fig. 1 p. 423).

The old male kouprey was carefully measured, photographed, skinned, and important parts of the skeleton as well as the skull were preserved. Through the kindness of Mr. Greenway and Mr. Edmond-Blanc this specimen procured by the Expedition in Indo-China was added to the share of the collections which were destined for this museum. Some of the photographs are reproduced here (see Plate II.)

Considerable local confusion has arisen over the fact that Cambodian natives are reported to distinguish two different kinds of wild cattle other than the gaur and the bantin. One they call the "Kouproh" and the other "Kouprey." Our specimen was locally called a "Kouproh" and appears to be indistinguishable from the "Kouprey" which was the form Urbain described as Bos (Bibos) sauveli. Edmond-Blanc is apparently of the same opinion and he writes on this subject in a private letter to Greenway from Paris on July 25, 1939: "Je crois en effet que le kouproh et le kouprey sont le même animal, mais suivant les régions les indigènes lui donnent l'un ou l'autre nom. Dans la région ou j'ai tué le mien, dans la province de Kratie près de Samrong, les Phnoms qui habitent la region appellent l'animal que j'ai tué kouproh mais disent qu'il existe sur le même territoire un autre animal qu'ils appellent kouprey et qui a l'avant-train identique au kouproh, mais l'arrière train du Bibos banteng, c'est-à-dire que l'arrière-train devient roux graduellement et qu'il a un disque fessier blanc. L'animal que vous avez à Boston est donc evidemment le même que celui que Urbain a décrit." In a letter received from the Compte de la Chevasnérie he says that Mr. Urbain and Mr. Omar Sarraut are of the same opinion as to the identity of this specimen with the kouprey. It is probably an older specimen than the few others that had previously been shot.

M. Urbain according to the article by Count de la Chevasnérie estimates the possible number of wild kouprey at about a thousand head. Our only definite information on localities where they exist is shown on the map in the introduction. These are Tchep, type locality, and Samrong near Krati where our specimen was procured, and where Pietri must have shot the specimen mentioned by Mr. Richmond. Probably that of Mr. Cornell came from the same locality as Pietri was his guide. The fact that the probable kouprey range is greater than the area designated encourages one to hope that they may be more numerous than present estimates would indicate. It is nevertheless most essential that the Government in Indo-China should immediately recognize the importance of preserving the kouprey only allowing specimens to be shot on a scientific permit!

BIBLIOGRAPHY FOR THE KOUPREY

1930. Defosse, Chasse et tourisme au Cambodge et dans le Sud Indo-Chine, *Société des éditions d'Extrême Asie*, Saigon.

1933. Vittoz, R., Étude zootechnique de l'élévage et de l'exploitation des bovins du Sud Indochinois en Cochinchine, *Bulletin economique de l'Indochine*, p. 947.

1937. Urbain, Ach., Le Kou-Prey ou Boeuf Gris Cambodgien, *Bull. Soc. Zool. de France*, pp. 305–307.

1937. Urbain, Ach., Le Kou-Prey ou Boeuf Sauvage Cambodgien, *Mammalia*, pp. 256–258.

1938. Urbain, Ach., Le Kou-Prey ou Boeuf Sauvage Cambodgien, *Bijdragen tot de Dierkunde, Feestnummer*, Amsterdam, pp. 545 & 546.

1939. Urbain, Ach., Rode, P. et Pasquier, M-A., La collection des Bovines asiatiques du Parc Zoologique du Bois de Vincennes, *Mammalia*, pp. 122-125.

1939. Urbain, Ach. Note complémentaire sur le Boeuf Sauvage du Cambodge (Bos (Bibos) sauveli Urbain), *Bulletin du Muséum*, 2e s., t. XI, n° 6.

1939. Sarraut, Omer, The Kouprey, *Big-Game Hunting in French Indo-China*, G. Taupin et Cie, Hanoi, pp. 16 & 17.

GENERAL BIBLIOGRAPHY

ADAMETZ, L.

1894. Untersuchungen über das Rind der Wahima-(Watussi-)Stämme. Bos (Zebu) africanus Watussi.) Journal für Landwirtschaft, Berlin, Jahrgang 42, pp. 137–155.

1898. Studien über Bos (brachyceros) europaeus, die wilde Stammform der Brachyceros-Rassen des europäischen Hausrindes. Journal für Landwirtschaft, Berlin, Jahrgang 46, pp. 269–320.

1898. Bos (brachyceros) europaeus, n. sp. Bull. internat. Acad. Cracovie, p. 88.

1930. Der sexuale Dimorphismus am Schädel des Urus u. seine Beziehungen zum Rassen- u. Abstammungsproblem des Hausrindes. Biologia generalis, vi, p. 1.

ANONYMOUS

1930. Le Temple d'Angkor Vat, Tome II; La Sculpture Ornamentale du Temple, Deuxième Partie, Mémoires Archéologiques, L'Ecole Française d'Extrême-Orient, Paris, plates 168 & 184.

AUERBACH, M.

1907. Auerochs und Wisent in Deutschland. . . .Verhandl. naturwis. Vereins in Karlsruhe, xx, p. 3.

BAUME, W. LA,

1909. Beiträge zür Kenntnis der fossilen u. subfossilen Boviden. Schrift. naturf. Ges. Danzig (II), xii, p. 45.

BLANFORD, W. T.

1888–91. Mammalia, The Fauna of British India including Ceylon and Burma. London, Taylor and Francis, pp. 482–493.

BOJANUS, L. H.

1826. De Uro nostrate eiusque skeleto. . . , Nova Acta . . . Acad. Caes. Leopold. Naturae Cur., xiii, p. 411.

BOULE, M.

1910. Les Grottes de Grimaldi, i, fas. 3, pp. 232–6 (Monaco).

1928. Breuil, H., Licent, E., et Teilhard, de Chadin P. Le Paléolithique de la Chine. Arch. Inst. Paléont. Humaine, Mém. 4, p. 78.

BRANDT, J. F.

1867. Zoogeographische u. paläontologische Beiträge, 2te Abhandlung, Die geog. Verbreitung des Zebu oder Bison, des Auerochsen der Neuern (Bos bison seu bonasus), p. 101; 3te Abhandlung, Die geog. Verbreitung des Ur oder wahren Auerochsen Bos primigenius seu Bos taurus sylvestris), p. 153. A brief abstract is given in the Quart. Journ. Geol. Soc, xxiii, p. 15.

CLARK, J. W.

1874. On a Nearly Complete Skeleton of Bos primigenius found in Burwell Fen. Proc. Camb. Phil. Soc., ii, p. 357.

COLBERT, E. H.

1935. Distributional and phylogenetic studies on Indian fossil Mammals. I. American Museum collecting localities in Northern India. Amer. Mus. Nov. no. 796, pp. 1–20, 12 text figs.

1935. Siwalik Mammals in the American Museum of Natural History. Trans. Amer. Phil. Soc., N.S., xxvi, pp. 401, map and 198 text figs.

CUVIER, G.

1823. Recherches sur les Oss. foss. 3me ed., Pt. iv, p. 150.

DAWKINS, WILLIAM BOYD.

1866. On the fossil British oxen. Part 1. Bos Urus Caesar. Quarterly Journal of the Geological Society of London. Proceedings. London, vol. 22, pp. 391–402.

1867. On the British Fossil Oxen. Pt. ii. Bos longifrons Owen. Quart. Journ. Geol. Soc., xxiii, p. 176.

DAWKINS, WILLIAM BOYD AND AISHFORD, SANFORD W.

1866–87. British Pleistocene Mammalia. Parts 1–6. London.

DÜRST, J. M.

1903. Quelques Bovidés préhistoriques. L'Anthropologie, ci, pp. 129 & 655.

1909. Animal Remains from the Excavations at Anau (Turkestan). Carnegie Institution of Washington, no. 73, p. 361.

DÜRST, J. ULRICH

1899. Die Rinder von Babylon, Assyrien und Ägypten und ihr Zusammenhang mit den Rindern der alten Welt. . . Berlin, 94 pp.

1900. Notes sur quelques Bovidés Préhistoriques, Mémoires Originaux. L'Anthropologie, Tome, Paris, Masson et Cie, pp. 129–158, & pp. 655–676.

1908. Animal remains from the excavations at Anau, and the horse of Anau in its relation to the races of domestic horses. Explorations in Turkestan. Expedition of 1904, Washington, vol. 2, pp. 341–344.

1926. Das Horn der Cavicornia, Seine Entstehungsursache, seine Entwicklung, Gestaltung and Einwirkung auf den Schädel der horntragenden Wiederkauer, Eine Monographie der Horner, Band 63, Abh. 1, Denkschriften der Schweizerischen Naturforschenden Gesellschaft.

EWART, J. COSSAR

1909. Hornless cattle. Live Stock Journal, London, vol. 70, Dec., pp. 599–600.

1911. On skulls of oxen from the Roman military station at Newstead, Melrose. Proc. Zool. Soc. of London, part 2, pp. 249–282.

1912. The Principles of Breeding and the Origin of Domesticated Breeds of Animals. 26th Annual Report of the Bureau of Animal Industry, for the year 1910, Washington, U. S. Department of Agriculture, pp. 125–186.

FALCONER, HUGH

1859. Descriptive Catalogue of the Fossil Remains of Vertebrata from the Siwalik Hills, the Nerbudda, Perim Island, etc. in the Museum of the Asiatic Society of Bengal, Calcutta. 261 pp.

1868. Palaeontological Memoirs and Notes (edited by C. Murchison), p. 280–289.

FALCONER, H., AND WALKER, H.

1859. Catalogue of Fossil Vertebrata in the Museum of the Asiatic Society of Bengal, Calcutta, 8vo., pp. 1–261.

FLOWER, W. H., AND LYDEKKER, R.

1891. Mammals living and extinct. London, 8vo., pp. xvi, 763, 357 text figs.

FOSSE, W. DE LA

1929. Contribution à l'étude de la faune quaternaire du Département de la Moselle. Le Bos primigenius Boj. Bull. Soc. Hist. Nat. Moselle (32), viii, p. 62.

FIEDLER, H.

1907. Über Säugetierreste aus braunschweigischen Torfmooren, nebst einem Beitrag zur Kenntniss der osteologischen Feschlechtscharactere des Rindschädels. Zeitsch. für Ethnologie, xxxix, p. 449.

FRANTZIUS, A. V.

1878. Die Urheimath des europäischen Hausrindes. Archiv für Anthropologie, Band 10, Braunschweig, pp. 129–137.

FRIZZELL, DONALD LESLIE

1933. Terminology of Types. The American Midland Naturalist, xiv, no. 6, p. 653.

GROMOVA, V.

1931. Contribution à la connaissance de l'Ure (Bos primigenius Boj.) de l'Europe orientale et de l'Asie septentrionale. Ann. Mus. Zool. Acad. Leningrad, xxxii, p. 293 (Russian).

HARTMANN, ROBERT

1864. Die Haussäugethiere der Nilländer. Annalen der Landwirthschaft in den Königlich Preussischen Staaten, Berlin, Band 43, pp. 281–310; Band 44, pp. 7-38.

1869. Studien zur Geschichte der Hausthiere. Zeitschrift für Ethnologie. . . . Band 1, Berlin, pp. 66, 232, 353.

HUBBACK, THEODORE

1938. Principles of Wild Life Conservation, The Journal of the Bombay Natural History Society, Bombay, vol. xl, no. 1, pp. 100–111, 4 plates.

1939. Wild Life Photography in the Malayan Jungle, The Journal of the Bombay Natural History Society, Bombay, vol. xli, no. 1, pp. 48–63, 8 plates.

HUGHES, T. MCKENNY

1896. On the more important breeds of cattle which have been recognized in the British Isles in successive periods, and their relation to other archaeological and historical discoveries. Archaeologia, London, . . . (2nd series), vol. 5, pp. 125–158.

JOHNSTON, H. H.

1903. British Mammals, London.

KELLER, CONRAD

1896. Das afrikanische Zebu-Rind und seine Beziehungen zum europäischen Brachyceros-Rind. Vierteljahrsschrift der Naturforschenden Gesellschaft in Zurich, pp. 455–487.

1897. Die Afrikanischen Elemente in der europaischen Haustierwelt. Globus, Band 72, Braunschweig, Nov. 6, pp. 285–289.

1898. Nochmals die Goldbecher von Vaphio. Globus, Braunschweig, vol. 74, pp. 81–82.

1899. Die Haustierwelt Asiens. Westermann's Illustrierte Deutsche Monatshefte . . . Jahrgang 43, Braunschweig, Band 86, pp. 567–575.

1902. Die Abstammung der altesten Haustiere. Zürich Fritz Amberger vorm, David Bürkli, pp. 232.

1905. Naturgeschichte der Haustiere. Berlin. 304 pp.

1909. Die Antike Tierwelt. Erster Band, Leipzig, Verlag von Wilhelm Engelmann, pp. 329–372.

1911. Studien über die Haustiere der Mittelmeer-Inseln. Ein Beitrag zur Losung der Frage nach der Herkunft der europaischen Haustierwelt. Band. xlvi, Abh. 2, Neue Denkschriften der Schweizerischen Naturforschenden Gesellschaft, pp. 109–186, viii Tafelerklärungen.

KLAAUW, C. J. VAN DER

1931. The auditory bulla in some fossil Mammals with a general introduction to this region of the skull. Bull. Amer. Mus. Nat. Hist., lxii, pp. 1–352, 18 text figs.

KOLDEWEY, ROBERT

1925. Das Wieder Erstehende-Babylon, p. 43.

LÖNNBERG, E.

1933. Description of a fossil Buffalo from East Africa. Arkiv. Zool. Stockholm, xxva, no. 17, pp. 1–32, 3 pls, 2 text figs.

LYDEKKER, RICHARD

1877. Notices of New and other Vertebrata from Indian Tertiary and Secondary Rocks. Records Geol. Surv. India, x, p. 30.

1880. Crania of Ruminants from the Indian Tertiaries. Memoirs of the Geological Survey of India. Palaeontologia Indica, series 10, vol. 1, pp. 88–181. Calcutta, pp. 88–153, pls. xi–xxiv.

1885. Catalogue of the Fossil Mammalia in the British Museum, pt. ii, London, pp. 2-38.

1877, 1883, 1887. (Fossil Vertebrata of India) Records of the Geological Survey of India, Calcutta, vol. 10, p. 30; vol. 16, p. 61; vol. 20, p. 51.

1898. Wild Oxen, Sheep and Goats of all Lands, Living and Extinct (Rowland Ward, London).

1912. The Ox and its Kindred. Methuen & Co., London, xi, 271 pp.

1913. Catalogue of the Ungulate Mammals in the British Museum. Vol. I. London, pp. 9-72.

1924. The Game Animals of India, Burma, Malaya, and Tibet. London, Rowland Ward, Ltd., pp. 50-89.

Matsumoto, H.

1915. On some fossil Mammals from Sze-Chuan, China. Sci. Rep. Tohoku Imp. Univ. (2) Geology, iii, pp. 1-28, pl. i-x, 4 text figs.

1918. On some fossil Bisontines of Eastern Asia. Sci. Rep. Tohoku Imp. Univ. (2) Geology, iii, pp. 83-102, pls. xxiv-xxxiv, 3 text figs.

Matthew, W. D., and Granger, W.

1923. New fossil Mammals from the Pliocene of Sze-Chuan, China. Bull. Amer. Mus. Nat. Hist., xlviii, pp. 563-598, 27 text figs.

Maxim, A.

1936. Bos primigenius Boj. din Quaternar . . . o privire sumara raspandirii lui in Romania (in Rumanian with abstract in German). Revista Mus Geologic Mineralogic al Universitatii din Clug. vi, p. 128.

Mertens, A.

1906. Der Ur, Bos primigenius Bojanus. Abh. u. Berichte Museum f. Natur u. Heimatkunde zu Magdeburg, p. 45.

Meyer, H. von

1835. Über fossile Reste von Ochsen, deren Arten u. das Vorkommen derselben. Nova Acta . . . Acad. Caes. Leopold. Naturae Cur., xvii, p. 100.

Middendorff, A. von

1888. Ueber die Rindviehrasse des nördlichen Russlands und ihre Veredlung. Landwirthschaftliche Jahrbücher, Berlin, Band 17, pp. 267-327.

Morin, Jean

1911. Les Animaux en Grèce, p. 127. Le Dessin des.

Morse, E. W.

1912. The Ancestry of Domesticated Cattle. U. S. Department of Agriculture, Bureau of Animal Industry, pp. 187-239.

Nehring, A.

1888. Ueber das Skelet eines weiblicher Bos primigenius aus einem Torfmoore der Provinz Brandenburg. Sitz. Ges. naturf. Freunde, Berlin, p. 54.

1888. Bos primigenius insbesondere über seine Coexistenz mit dem Menschen. Zeitsch. f. Ethnologie, xx.

1889. Über Riesen und Zwerge des Bos primigenius. Sitz. Ges. naturf. Freunde, Berlin, p. 5.

1892. Über Atlas und Epistropheus des Bos primigenius. Sitz. Ges. naturf. Freunde, Berlin, p. 129.

1896. Die Herberstain, schen Abbildungen des Ur und des Bison, Landwirtsch. Jahrb., xxv, p. 915.

1897. Ueber Herberstain und Hirsfogel . . . Berlin.

Nilsson, S.

1849. On the Extinct and Existing Bovine Animals of Scandinavia. Ann. and Mag. Nat. Hist. (2) iv, pp. 256–69 and 349–55.

Oliver, Col. Sir Arthur

1938. A Brief Survey of Some of the Important Breeds of Cattle in India. Miscellaneous Bulletin, No. 17, The Imperial Council of Agricultural Research, Delhi, pp. 45, plates LI.

Osgood, Wilfred H.

1932. Mammals of the Kelley-Roosevelt and Delacour Asiatic Expeditions. Field Museum of Natural History Publication 312, Zoological Series, vol. xviii, No. 10, pp. 337 & 338.

Owen, Richard

1843. Report on the British Fossil Mammalia, Pt. II, Ungulata, Rep. Brit. Assoc. (Cork), p. 232.

1846. A History of British Fossil Mammals and Birds. London, John Van Voorst, pp. xlvi, 560.

1860. Paleontology, or a systematic summary of extinct animals and their geological relations. Edinburgh.

Pagenstecher

1878. Studien zum Ursprung des Rindes, mit einer Beschreibung der fossilen Rinderreste des Heidelberger Museums. Fühling's Landwirthschaftliche Zeitung, Jahrgang 27, Berlin and Leipzig.

Pallas, P. S.

1769. De Ossibus Siberiae Fossilibus. . . . Novi Comment. Acad. Scient. Imp. Petropol., xiii, p. 460.

Parr, Virgil V.

1924. Brahman (Zebu) Cattle. U. S. Department of Agriculture, Farmers' Bulletin No. 1361, pp. 21.

PILGRIM, G. E.

1907. Note on a cranium of Boselaphus namadicus Rütim. from the Narbada Pleistocene gravels of Jabalpur. Rec. Geol. Surv. Ind., xxxv, pp. 120–121.

1937. Siwalik Antelopes and Oxen in the American Museum of Natural History. Bulletin of the American Museum of Natural History, lxxii, Art. vii, pp. 815–825.

1939. The Fossil Bovidae of India. Mem. Geol. Surv. Ind., Palaeontologia Indica (N.S.), xxvi, Mem. I pp. 365, pls. i–viii.

PILGRIM, G. E., AND HOPWOOD, A. T.

1928. Catalogue of the Pontian Bovidae of Europe. British Museum (Natural History), pp. viii, 106, pls. i–ix, 3 text figs., London.

POCOCK, R. I.

1910. On the specialized cutaneous glands of Ruminants. Proc. Zool. Soc. London, pp. 840–986, 142 text figs.

1918–19. On some external characters of Ruminant Artiodactyla. Ann. Mag. Nat. Hist., (9) iii, pp. 103–114.

POHLIG, H.

1911. Bovidés fossiles de l'Italie. Bull. Soc. belge Géol. Pal. Hydrol., xxv, Procès-verb., p. 311.

POMEL, A.

1894. Boeufs-Taureaux. Paleontol. Monograph. Carte géol. de l'Algérie, p. 5.

PUMPELLY, RAPHAEL

1908. Prehistoric Civilizations of Anau, Origins, Growth, and Influence of Environment. Vol. II, Explorations in Turkestan, Washington, D. C., Carnegie Institution, pp. 359–400.

REINECKE, T. G. W.

1924. Africander Cattle. Department of Agriculture, Union of South Africa, Pretoria, pp. 15.

REYNOLDS, SIDNEY H.

1939. British Pleistocene Mammalia. Vol. III, Part VI, The Bovidae, London, Palaeontographical Society, pp. 65.

RITCHIE, J.

1920. The Influence of Man on Animal Life in Scotland, a Study in Faunal Evolution.

ROMER, A. S.

1928. Pleistocene Mammals of Algeria, Fauna of the Palaeolithic Station of Mechta-el-Arbi. Logan Mus. Bull, i, no. 2, published Beloit, Wis., p. 79.

1938. Part II, Mammalian Remains from Some Paleolithic Stations in Algeria. Logan Museum Bull. v, pp. 170–172.

ROSTAFINSKI, J.

1933. Proba systemat. Malych Bovidow Europy (in Polish with an English summary). Rozpr. Biolog. Med. Wet. Rol. i Hod., t. xi; z. 3, Lwow, p. 1.

RÜTIMEYER, L.

1862. Die Fauna der Pfahlbauten in der Schweiz. Neue Denkschriften der allgemeinen Schweizerischen Gesellschaft für die gesammten Naturwissenschaften. Zürich, Band 19, pp. 1–248.

1867. Versuch einer natürlichen Geschichte des Rindes in seinen Beziehungen zu den Wiederkäuern im Allegemeinen. Zürich, pp. 102–175.

1877–78. Die Rinder der Tertiär-Epoch nebst Vorstudium zu einer natürlichen Geschichte der Antilopen. Abh. Schweiz. Paläon. Ges. Zürich, pls. vii, 8 text figs., pp. 113–203.

SCHWARZ, ERNST

1913. Die Fossilen Antilopen von Oldoway. Wissenschaftliche Ergebnisse der Oldoway-Expedition 1913, Berlin. pp. 13–25.

SIEGFRIED, H.

1907. Die Rinderschädelfunde von Pasquart u. deren Stellung zu den subfossilen u. rezenten Rinderrassen. Abh. Schweiz. Päl. Ges., xxiv, p. 1 (Basel).

SISSON, SEPTIMUS T.

1917. The Anatomy of the Domestic Animals. Philadelphia and London, W. B. Saunders Company, pp. 930.

STAUDINGER, W.

1909. Vergleichende Untersuchungen am Skelette der quartaren u. rezenten Wildrinder Europas. Beiträge zur Urgeschichte der Hausrinder, Halle.

STEFANO, G. DE

1913. . . . fossili del genere Bos Linneo . . . dell isola di Pianosa. Boll. Soc. geol. ital., xxxii, p. 49.

1928. Kleiner Bovide. Eclogae Geologicae Helvetiae, Tome 21, Bâle, Emile Birkhaeuser & Cie, p. 441f.

TRÖLTSCH, E. VON

1902. Die Pfahlbauten des Bodenseegebietes, Stuttgart, 255 pp.

WEIGALL, ARTHUR

1924. Ancient Egyptian Works of Art. London, T. Fisher Unwin, Ltd., p. 135 & 329.

WERNER, HUGO

1892. Die Rinderzucht . . . 645 pp. 136 plates. Berlin, 1892. Second edition, 1902.

WILCKENS, MART

1876. Die abändernden Einflüsse der Kultur auf die Form des Rinderschädels. Landwirthschaftliche Jahrbücher, Band 5, Berlin. pp. 651–653.

1877, 78, 80. Ueber die Schädelknochen des Rindes aus dem Pfahlbau des Laibacher Moores. Mittheilungen der Anthropologischen Gesellschaft in Wien, Band 7, pp. 165–175. Wien, Band 9.

1885. Grundzüge einer Naturgeschichte des Hausrindes. Vienna.

1885. Die Abstammung der Rindes und die tertiären Formen desselben. Biol. Centralbl, iv, p. 749.

1885. Zur Geschichte des europäischen Urochsen. Landwirtsch. Jahrb., xiv, Berlin. p. 263.

1905. Naturgeschichte der Haustiere. New edition edited by J. M. Duerst (Leipzig),

1905. Grundzüge der Naturgeschichte der Haustiere. 2nd edition, neubearbeitet von Duerst, J. U., Leipzig, 8vo., pp. i–xi, 408, 85 text figs.

WINGE, H.

1924. Pattedyr-Slaegter, III, Ungulata og Cetacea, Copenhagen, 8vo., pp. 1–170.

WOODWARD, A. S.

1924. Some Remarks on the Pleistocene Mammalia. Essex Naturalist, xxi, p. 1.

WOODWARD, A. SMITH, AND SHERBORN, C. DAVIES

1890. A Catalogue of British Fossil Vertebrata (Dulau, London).

ZERVOS, CHRISTIAN

1935. L'Art de la Mesapotamie, p. 66.

EXPLANATION OF THE PLATES

PLATE 1

PLATE 1

Novibos sauveli, (Urbain). An old male kouprey specimen now in the collection of the Museum of Comparative Zoology (M.C.Z. 38108), first described in this memoir. This specimen was shot by Mr. François Edmond-Blanc in March, 1939, and presented to the Museum by James C. Greenway, Jr. $\times{}^{1}/_{12}+$.

PLATE 2

PLATE 2

Photographs of the kouprey (M.C.Z. 38108) taken by Mr. François Edmond-Blanc in the field near Samrong in Cambodia just after the animal was shot.

Fig. 1. Note the open savannah forest, which is the characteristic habitat of the kouprey.

Fig. 2. Note the shape of the face and the heavy wrinkling of the skin; also the shape of the ear and nostril.

Fig. 3. Note the shape of the nostril and the spread of the horns which are frayed near the tips.

Fig. 4. Note the approximation of the bases of the horns and the angle at which they leave the skull.

1

2

3

4

PLATE 3

PLATE 3

Fig. 1. A silhouette photograph of the holotype of *Novibos sauveli*, (Urbain). This picture was taken by Mr. François Edmond-Blanc of the 3½ year old male kouprey at the Vincennes Zoo in June, 1939.

Fig. 2. Occipital view of the kouprey skull (M.C.Z. 38108) showing the prominent posterior primary keel on the horn-core and the characteristic fraying of the horn sheath.

Fig. 3. The kouprey frontlet (A.M.N.H. 89003) loaned by the American Museum of Natural History, and shot by Mr. Ezra B. Cornell of Rumford, Rhode Island.

1

2

3

PLATE 4

PLATE 4

Occipital view of skull of *Novibos sauveli* (M.C.Z. 38108). Note the cross sections of the horn-core indicating the primary posterior keel. These sections were made at 100 mm. intervals. × 2/5.

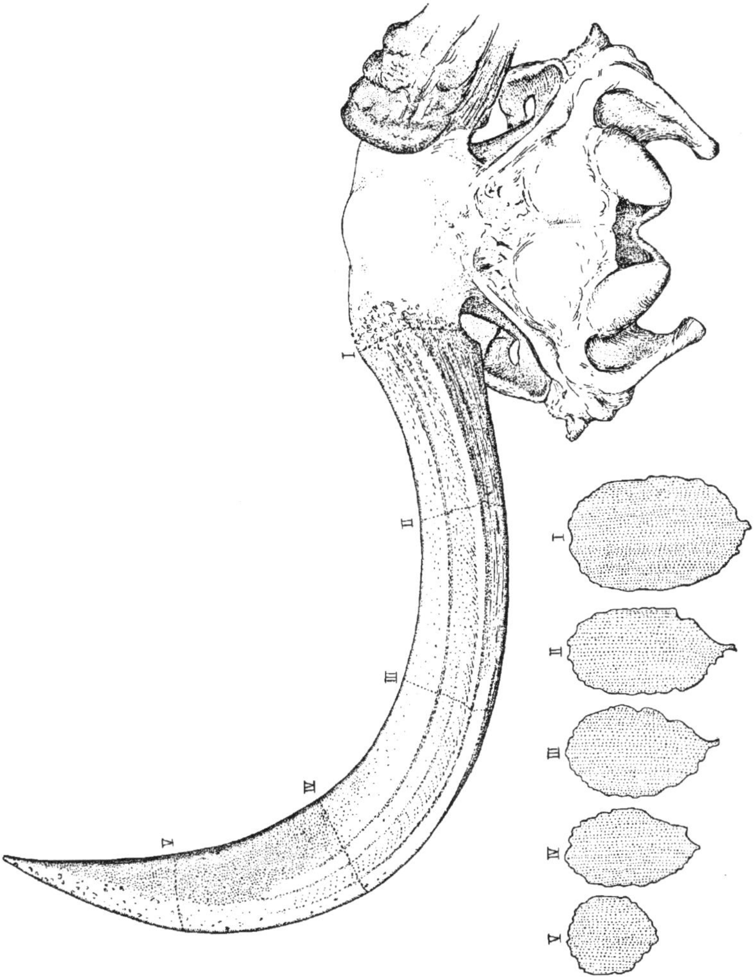
I
II
III
IV
V
I
II
III
IV
V

PLATE 5

PLATE 5

Fig. 1. Kouprey frontlet (A.M.N.H. 89003). Basal view of kouprey frontlet (A.M.N.H. 89003) showing the shape of the top of the brain case, the suture line of the parietal and frontal, and the marked thickening of the occiput with its sinuses. $\times ^2/_7$.

Fig. 2. Dorsal view of the frontlet showing the heavy ridging at the base of the horns and the characteristic fraying of the horn sheath, which is coarser in this animal than in the older M.C.Z. kouprey. $\times ^2/_7$.

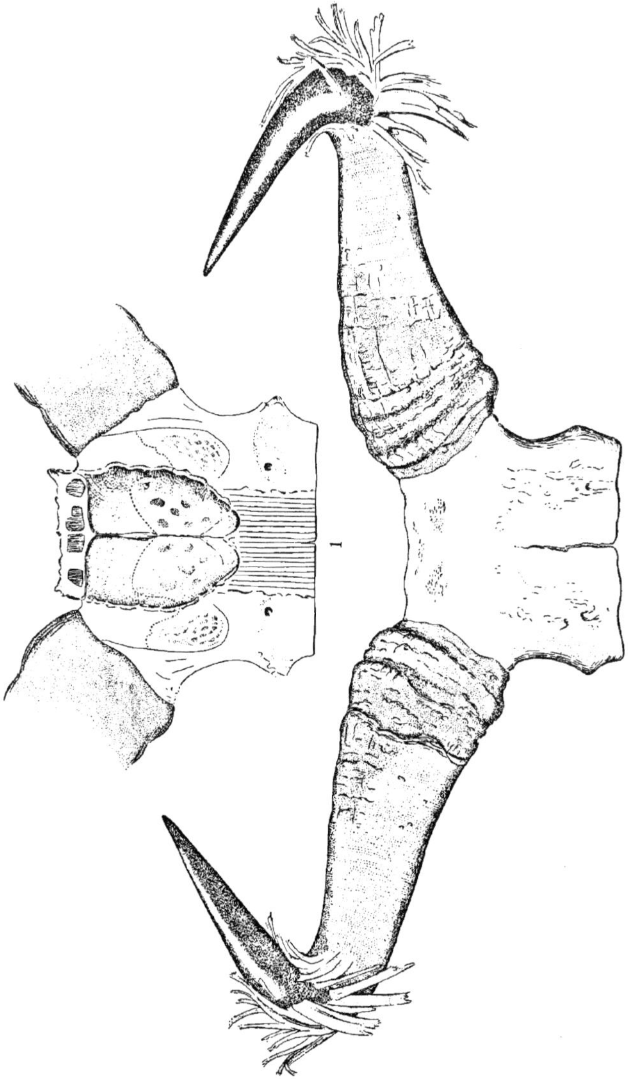
1
2

PLATE 6

PLATE 6

Lateral view of the skull of *Novibos sauveli* (M.C.Z. 38108). $\times \frac{3}{8}$.

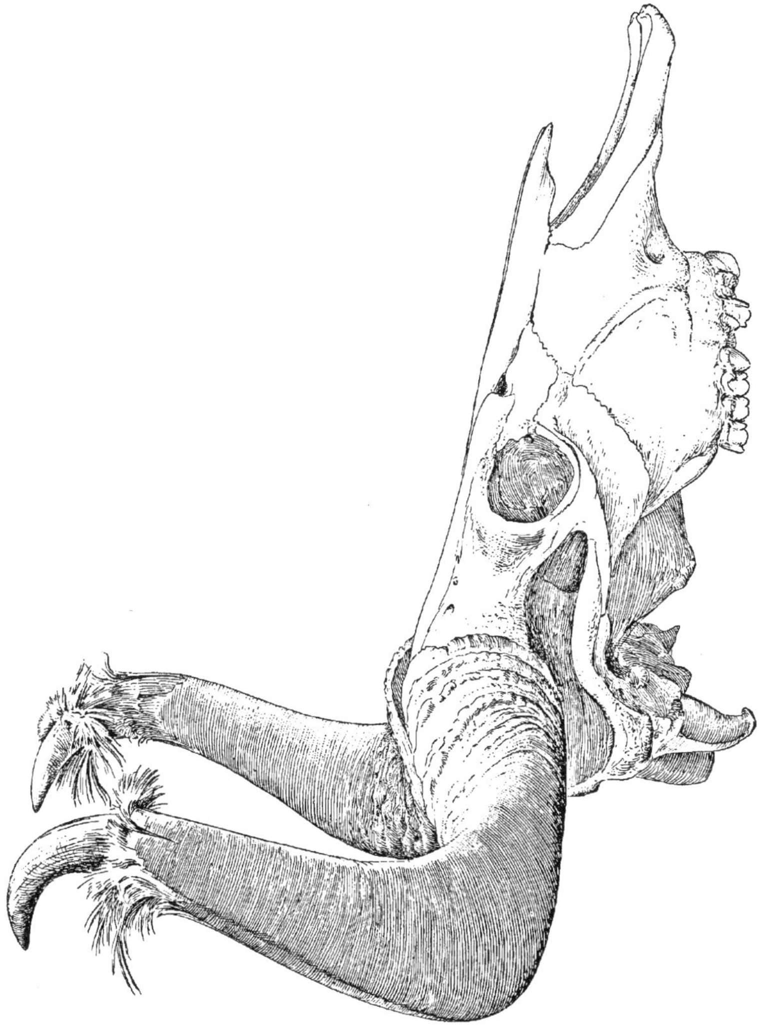

PLATE 7

PLATE 7

Dorsal view of the skull of *Novibos sauveli* (M.C.Z. 38108). ×⅓.

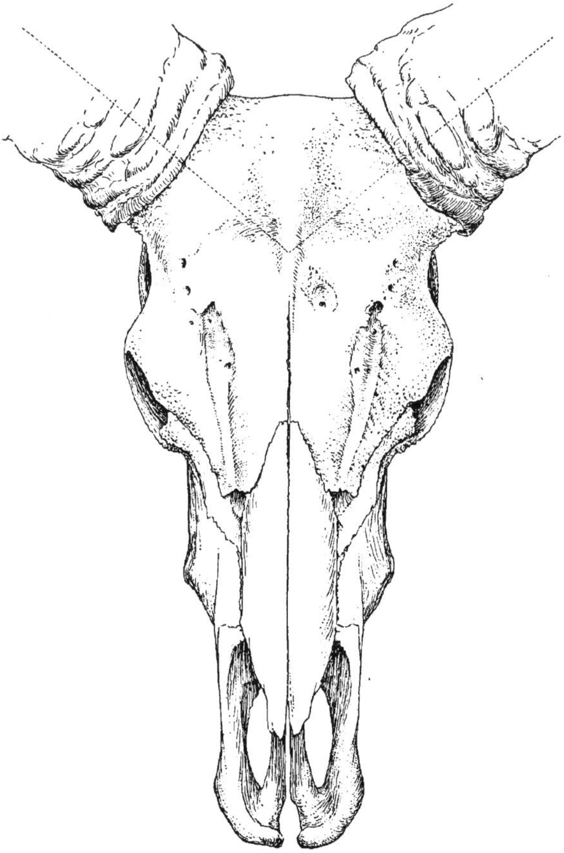

PLATE 8

PLATE 8

Basal view of the skull of *Novibos sauveli* (M.C.Z. 38108). × ⅓.

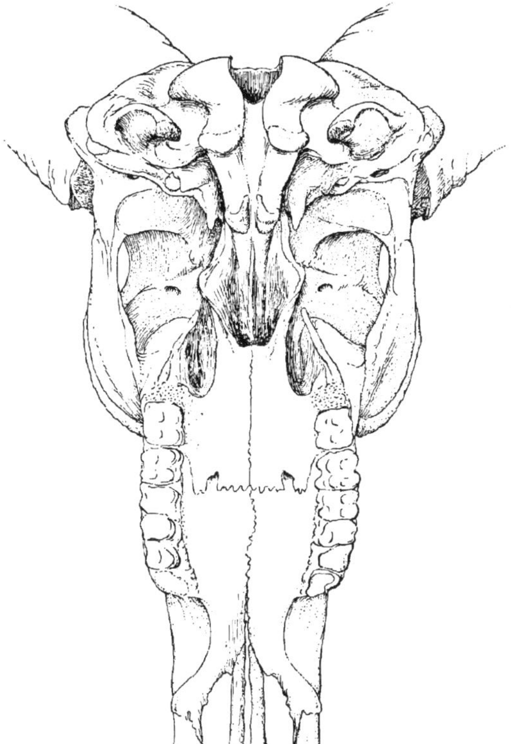

PLATE 9

PLATE 9

Basal view of the skull of *Bibos banteng* subsp? (M.C.Z. 36669). × ⅓.

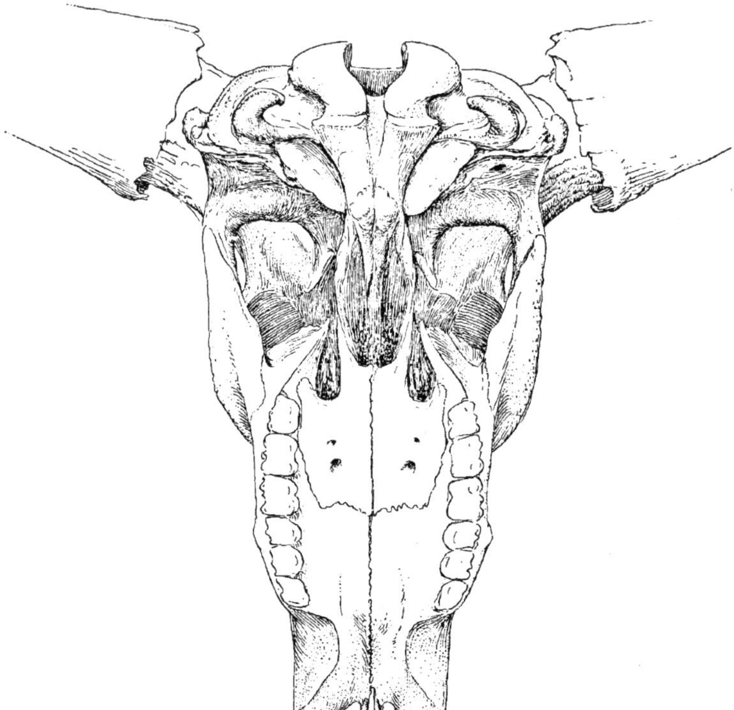

PLATE 10

PLATE 10

Basal view of the skull of *Bibos gaurus readi* (M.C.Z. 36670). × ⅓.

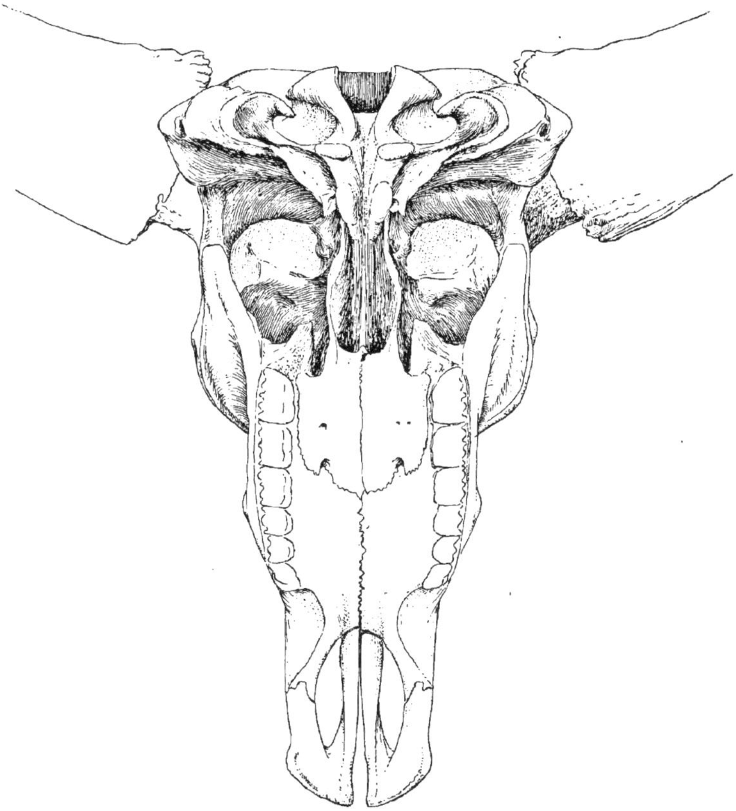

PLATE 11

PLATE 11

Comparative nuchal views. ×¼.

Fig. 1. *Bibos banteng* Subsp? (M.C.Z. 36669).

Fig. 2. *Novibos sauveli* (M.C.Z. 38108).

Fig. 3. *Bibos gaurus readi* (M.C.Z. 36670).

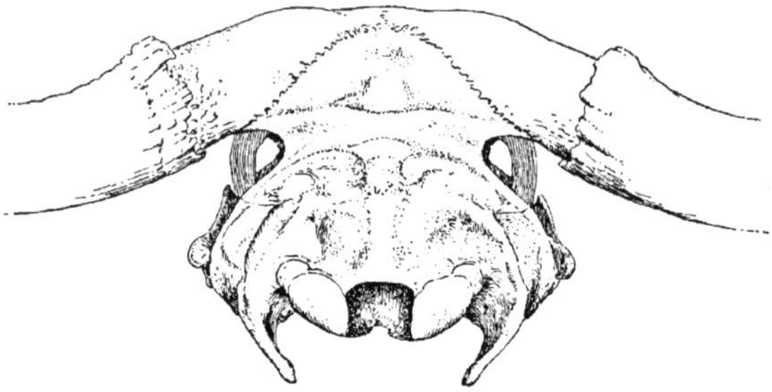

1

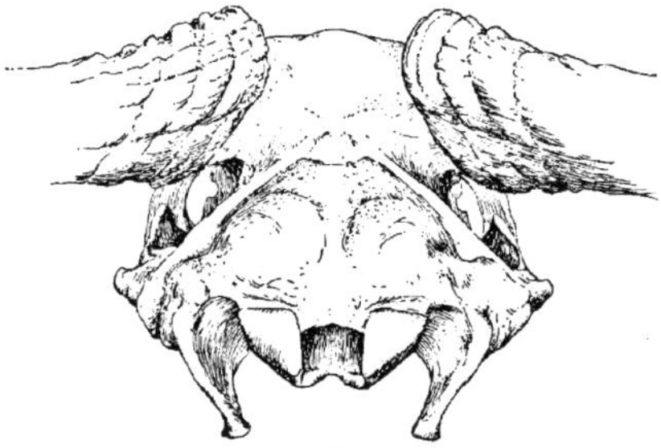

2

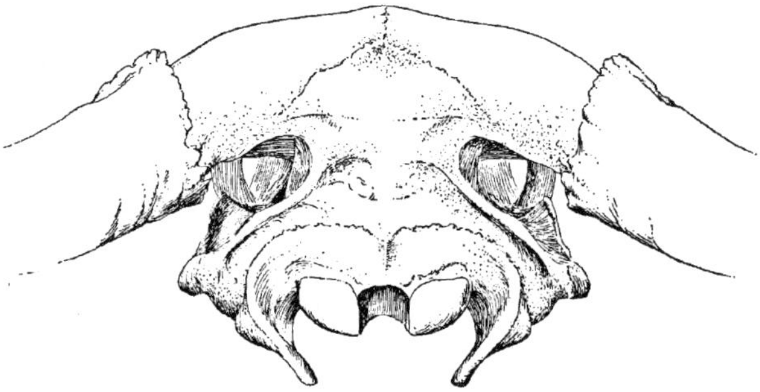

3

Spotted something that isn't quite right?

Let us know and we'll fix it for future reprints. And as a thank you for your diligence, we'll send you a free Sothis book of your choice.

publisher@sothis.press

This page intentionally left blank

Made in the USA
Las Vegas, NV
03 May 2026

46294358R10090